Mathematical Engineering

Today, the development of high-tech systems is unthinkable without mathematical modeling and analysis of system behavior. As such, many fields in the modern engineering sciences (e.g. control engineering, communications engineering, mechanical engineering, and robotics) call for sophisticated mathematical methods in order to solve the tasks at hand.

The series Mathematical Engineering presents new or heretofore little-known methods to support engineers in finding suitable answers to their questions, presenting those methods in such manner as to make them ideally comprehensible and applicable in practice.

Therefore, the primary focus is—without neglecting mathematical accuracy—on comprehensibility and real-world applicability.

To submit a proposal or request further information, please use the PDF Proposal Form or contact directly: Dr. Thomas Ditzinger (thomas.ditzinger@springer.com)

Indexed by SCOPUS, zbMATH, SCImago.

More information about this series at http://www.springer.com/series/8445

Tofigh Allahviranloo · Armin Esfandiari

A Course on Integral Equations with Numerical Analysis

Advanced Numerical Analysis

Springer

Tofigh Allahviranloo
Faculty of Engineering and Natural Sciences
Bahcesehir University
Istanbul, Turkey

Armin Esfandiari
Faculty of Engineering and Natural Sciences
Bahcesehir University
Istanbul, Turkey

ISSN 2192-4732 ISSN 2192-4740 (electronic)
Mathematical Engineering
ISBN 978-3-030-85349-5 ISBN 978-3-030-85350-1 (eBook)
https://doi.org/10.1007/978-3-030-85350-1

This Springer imprint is published by the registered company Springer Nature Switzerland AG
The registered company address is: Gewerbestrasse 11, 6330 Cham, Switzerland

Preface

As the title of the book suggests, the numerical analysis subjects' matter are the important tools of the book topic. Because numerical errors and methods have important roles in solving integral equations. Therefore, all needed topics including a brief description of interpolation are explained in the book.

The integral equations have many applications in the engineering, medical and economic sciences, so the present book contains new and useful materials about interval computations including interval interpolations that are going to be used in interval integral equations.

The concepts of integral equations are going to be discussed in two directions, analytical concepts and numerical solutions which both are necessary for these kinds of dynamic systems. The differences between this book with the others are a full discussion of error topics and also using interval interpolations concepts to obtain interval integral equations.

All researchers and students in the field of mathematical, computer and also engineering sciences can benefit the subjects of the book.

Istanbul, Turkey

Tofigh Allahviranloo
Armin Esfandiari

Contents

Chapter 1
Introduction to Numerical Analysis

1.1 Introduction

Bases on the discussed numerical methods and considering other chapters of the book, we need to explain the error analysis first. In addition to explore the numerical integration methods, the interpolation of integrand function is needed, and this is one of the reasons to discuss about interpolation methods.

1.2 Error Analysis

In this chapter, we intend to investigate and analyze the important complicated problems and points that occur in numerical calculations or calculations based on the numerical algorithms. As we know, in the numerical analysis, most numerical methods are iterative. This means that their formulation is in the form of difference equations. Therefore, given one or more initial values, the next values must be calculated. Usually, the initial values are not accurately available and are approximate; or due to the structure of the mathematical model, the calculations performed using iterative methods produce approximate results, that is, whether the initial values are approximate or not, the difference model may also have error factors. Obviously, two points were always considered in the computer or the numerical calculations. One is the speed of calculations and the other is the memory occupied by numerical results. Due to the advances in science and technology, the second factor has been ignored in the presentation of structured algorithms, but the first factor is considered as an advantage for the presentation of numerical algorithms. Given that each computational device has its own computational accuracy, it can be said that the zero of each computing device or computer is different from that of another computer, that is, the smallest positive number of one machine is different from that of another machine. Therefore, an algorithm performed in two machines will have different results! But in both cases, there is a computational error which is less in one than the

T. Allahviranloo and A. Esfandiari, *A Course on Integral Equations with Numerical Analysis*, Mathematical Engineering, https://doi.org/10.1007/978-3-030-85350-1_1

other. Currently, due to the advances in technology and the construction of advanced satellites and long-range air-to-air missiles and missiles with nuclear warheads, an approximate estimate of the target with the lowest error rate and the calculations of missile or satellite launch with the least amount of error is important. This is because the missile is trying to hit a specified target over a distance of, for example, thousands of kilometers, which may also be approximate. However, how to launch the missile, initial speed, initial acceleration, traveled distance, obstacles in the path of the missile such as air resistance, winds blowing from lateral directions, etc., and how to hit the target, all are factors that required to be considered, and obviously none of these factors can be accessed accurately and without an error. Therefore, taking into account these factors and problems, hitting the target with a missile should be done with an error of, for example, a maximum of 0.01. Obviously, all models related to this process are in the form of mathematical models, for example, the differential equations with the initial conditions, the integral equations, the differential equations with partial derivatives, the calculations of integral series and so on. So we need to examine the errors of such models and estimate the upper and lower bounds of such errors. In this regard, some problems about error analysis are presented.

1.2.1 Errors in an Algorithm

Suppose that $Y = \phi(X)$, where ϕ is a combination of all the steps of the algorithm. For this purpose, we define:

$$\phi : D \to \mathbb{R}^m$$

where D is an open subset of $\mathbb{R}^n$. We also assume that $X^t = (x_1, ..., x_n)$ and $Y^t = (y_1, ..., y_m)$ are the input and output vectors of the algorithm, respectively. It is clear that:

$$y_i = \varphi_i(x_1, ..., x_n), \, i = 1, ..., m$$

i.e.,

$$Y = \begin{bmatrix} y_1 \\ \vdots \\ y_m \end{bmatrix} = \begin{bmatrix} \varphi_1(x_1, ..., x_n) \\ \vdots \\ \varphi_m(x_1, ..., x_n) \end{bmatrix}$$

If we want to specify ϕ for an algorithm that has $r + 1$ operators (steps), we have:

$$\varphi^{(i)} : D_i \to D_{i+1}, \, i = 0, ..., r, \, D_i \subseteq \mathbb{R}^{n_i}, \, n_i \in \mathbb{Z}$$

$$\phi = \varphi^{(r)} \circ \ldots \circ \varphi^{(0)},\ D_0 = D,\ D_{r+1} \subseteq \mathbb{R}^{n_r+1} = \mathbb{R}^m$$

To calculate ϕ in each algorithm, we have the ordered series of $\varphi^{(i)}$ operators with the sum equal to ϕ, so that the output of one operator will be the input of the next operator, and finally the output of the last operator will be Y.

If in ith step of the algorithm, the vector $X^{(i)}$ has n_i inputs for the operator $\varphi^{(i)}$, then we have:

$$\varphi^{(i)} : D_i \to \mathbb{R}^{n_i+1},\ D_i \subseteq \mathbb{R}^{n_i}$$

So that

$$\varphi^{(i)}\left(X^{(i)}\right) = X^{(i+1)}$$

We have already mentioned that:

$$\phi : D \to \mathbb{R}^m,\ D \subseteq \mathbb{R}^n$$

$$\phi(X) = \begin{bmatrix} \varphi_1(x_1, \ldots, x_n) \\ \vdots \\ \varphi_m(x_1, \ldots, x_n) \end{bmatrix}$$

We assume that elements of $\phi(X)$, i.e., φ_is, have a continuous first derivative and $\tilde{X}$ is an approximation of x. We know that absolute error and relative error of $\tilde{X}$ are defined as follows:

$$\Delta\tilde{X} = \tilde{X} - X \tag{1.1}$$

$$\varepsilon_{\tilde{X}} = \frac{\tilde{X} - X}{X} \tag{1.2}$$

Considering the absolute error and relative error of $\tilde{X}$, the absolute error and the relative error of $\tilde{Y} = \phi\left(\tilde{X}\right)$ will be as follows:

$$\Delta\tilde{Y} \le \sum_{j=1}^{n} \left| \frac{\partial\varphi_i(X)}{\partial x_j} \right| .\Delta\tilde{x}_j \tag{1.3}$$

$$= D\phi(X).\Delta\tilde{X} \tag{1.4}$$

And by using the definition of relative error and formula (1.2)

$$\varepsilon_{\tilde{y}_i} \leq \sum_{j=1}^{n} \left| \frac{\partial \varphi_i(X)}{\partial x_j} \right| . \frac{x_j}{\varphi_i(X)} . \varepsilon_{\tilde{x}_j}$$

$$x_j \neq 0,\ j = 1, \ldots, n,\ y_i \neq 0,\ i = 1, \ldots, m \tag{1.5}$$

In relation (1.4), $D\phi(X)$ is the Jacobian matrix. The factors $\frac{\partial \varphi_i(X)}{\partial x_j}$ in (1.3) are called sensitivity numbers and also the factors $\frac{\partial \varphi_i(X)}{\partial x_j} . \frac{x_j}{\varphi_i(X)}$ in (1.5) are called condition numbers. If in a problem, all condition numbers are small, we say that the problem is well-established.

The following problem describes the analysis of the series calculation error with a computational tool. Due to the widespread application of series, it is important to explain such problems.

1.2.1.1 Problem

Calculating $\sum_{i=1}^{n} a_i$ results in an arbitrary large relative error. If all terms of a_i have the same sign, then the relative error is bounded, find a bound for this error, ignoring the higher-order terms.

Solution: Suppose that $y = \sum_{i=1}^{n} a_i$ and $\tilde{y}$ is an approximation of it. According to the relative error formula:

$$\varepsilon_{\tilde{y}} \leq \sum_{j=1}^{n} \frac{x_j}{\phi(x)} . \left| \frac{\partial \phi(x)}{\partial x_j} \right| . \varepsilon_{x_j}$$

we have:

$$\begin{aligned} \varepsilon_{\tilde{y}} &\leq \sum_{j=1}^{n} \frac{a_j}{a_1 + \cdots + a_n} . |1| . \varepsilon_{a_j} \\ &= \frac{a_1}{a_1 + \cdots + a_n} . \varepsilon_{a_1} + \frac{a_2}{a_1 + \cdots + a_n} . \varepsilon_{a_2} + \cdots + \frac{a_n}{a_1 + \cdots + a_n} . \varepsilon_{a_n} \end{aligned}$$

So:

$$\varepsilon_{\tilde{y}} \leq \left| \frac{a_1}{a_1 + \cdots + a_n} \right| . \left| \varepsilon_{a_1} \right| + \cdots + \left| \frac{a_n}{a_1 + \cdots + a_n} \right| . \left| \varepsilon_{a_n} \right|$$

If all terms have same sign, then:

$$\left| \frac{a_i}{a_1 + \cdots + a_n} \right| \leq 1,\ i = 1, \ldots, n$$

And if $\max_{i=1,\ldots,n} \left| \varepsilon_{a_i} \right| = M$, we will have:

$$\left|\varepsilon_{\tilde{y}}\right| \le \left|\varepsilon_{a_1}\right| + \cdots + \left|\varepsilon_{a_n}\right| \le n.M$$

So, the relative error has an upper bound.
Now if all terms have not same sign, then:

$$\exists i\left(i = 1, \ldots, n \Rightarrow \left|\frac{a_i}{a_1 + \cdots + a_n}\right| \ge 1\right)$$

and the error will be huge.

The purpose of expressing and solving the following problem is to point out that the subtraction of close approximate numbers of the same sign has a large relative error.

1.2.1.2 Problem

Consider a quadratic equation. If $b > 0$ and $b^2 \gg ab$, how should the root of the equation be calculated to minimize the calculation error?

Solution: The roots of the equation are:

$$x_1 = \frac{-b + \sqrt{b^2 - ac}}{a}, x_2 = \frac{-b - \sqrt{b^2 - ac}}{a}$$

Since $b > 0$ and $b^2 \gg ab$, so the value of $\sqrt{b^2 - ac}$ is very close to b, and therefore, in the nominator of x_1, we will have a subtraction of two close numbers, so there is a big error for calculating x_1, but in calculating x_2, there is not. So, first we obtain x_2 and then, we also calculate x_1 as follows:

$$\begin{aligned} x_1 &= \frac{-b + \sqrt{b^2 - ac}}{a} . \frac{-b - \sqrt{b^2 - ac}}{-b - \sqrt{b^2 - ac}} \\ &= \frac{b^2 - \left(b^2 - ac\right)}{a.x_2.a} \\ &= \frac{c}{ax_2} \end{aligned}$$

1.2.1.3 Definition

If $\tilde{x}$ is an approximation of x, then

$$e(\tilde{x}) = |x - \tilde{x}|$$

is called an absolute error of a_n.

1.2.1.4 Example

1. Suppose that $a_n = \frac{n+1}{n}$, what is the absolute error of a_n as an approximation of the number one?

$$e(a_n) = \left|1 - \frac{n+1}{n}\right| = \frac{1}{n}$$

It can be observed that the larger the n, the smaller the $\frac{1}{n}$, and as a result, the a_n gets closer to one. If we want the error of a_n to be smaller than, for example, 10^{-3}, it is enough to write:

$$\frac{1}{n} < 10^{-3}$$

Therefore, $n > 10^3$. The first n that holds in the recent inequality is 1001, for which:

$$a_n = \frac{1002}{1001} = 1.000999$$

2. We know that 1.141 is an approximation of $\sqrt{2}$. What is the absolute error of 1.41?

Obviously, if we write the decimal expansion of $\sqrt{2}$ using a calculator, i.e.,

$$\sqrt{2} = 1.414213562(9D)$$

then

$$e(1.41) = 0.0042135562000$$

$$2.66 \leq l \leq 2.74$$

Now the question is whether the absolute error of an approximation determines the accuracy of that approximation? Read the following and answer the relevant questions to find out the answer of the above question.

(A) Consider two bank cashiers, one of whom has lost one hundred Tomans by exchanging, for example, one million Tomans, and the other has extra one hundred Tomans by exchanging five hundred thousand Tomans!
Which cashier was more accurate?

(B) Consider two goalkeepers, one of whom has conceded 4 goals from 5 penalties and the other has conceded 6 goals from 10 penalties. Which goalkeeper was better?

From the above examples, it follows that what determines the accuracy of an approximation is the error per unit of quantity; the smaller the error, the accurate the approximation. Therefore, we have the following definition.

1.2.1.5 Definition—Relative Error

If $\tilde{x}$ is an approximation of a nonzero number x, the relative error of x is denoted by $\delta(\tilde{x})$, which is the error per unit of the quantity, i.e.,

$$\delta(\tilde{x}) = \frac{|x - \tilde{x}|}{|x|}$$

In most numerical analysis problems, there is not a real number x. For this purpose, an upper bound can be obtained for $\delta(\tilde{x})$ without the existence of x.

1.2.1.6 Problem

Suppose that $\bar{x}$ is an approximation of x and $\bar{y}$ is an approximation of y. Calculate the relative error $\frac{x}{y}$.

Solution: Suppose that $\varepsilon := \bar{x} - x$ and $\eta := \bar{y} - y$. So,

$$\begin{aligned}
\varepsilon_{\frac{x}{y}} &= \frac{\frac{\bar{x}}{\bar{y}} - \frac{x}{y}}{\frac{x}{y}} \\
&= \frac{\frac{x+\varepsilon}{y+\varepsilon} - \frac{x}{y}}{\frac{x}{y}} \\
&\simeq \frac{\frac{x(1+\varepsilon_{\bar{x}})}{y(1+\varepsilon_{\bar{y}})} - \frac{x}{y}}{\frac{x}{y}} \\
&= \frac{1+\varepsilon_{\bar{x}}}{1+\varepsilon_{\bar{y}}} - 1 \\
&\simeq \varepsilon_{\bar{x}} - \varepsilon_{\bar{y}}
\end{aligned}$$

1.2.1.7 Theorem

If $\tilde{x}$ is an approximation of x and $e_{\tilde{x}}$ is an absolute limit error of $\tilde{x}$, we have

$$\delta(\tilde{x}) \leq \frac{e_{\tilde{x}}}{|\tilde{x}| - e_{\tilde{x}}}$$

Proof According to the hypothesis, we have

$$|x - \tilde{x}| \leq e_{\tilde{x}}$$

and according to the properties of the absolute value,

$$|\tilde{x}| - |x| \leq |x - \tilde{x}|$$

As a result, the sentence is true.

If $e_{\tilde{x}}$ is small compared to $|\tilde{x}|$, it can be ignored and so we can write

$$|\tilde{x}| - e_{\tilde{x}} \simeq |\tilde{x}|$$

so, we have the following remark.

1.2.1.8 Remark

If $e_{\tilde{x}}$ is small compared to $|\tilde{x}|$

$$\delta(\tilde{x}) \lesssim \frac{e_{\tilde{x}}}{|\tilde{x}|}$$

Practically, the right side of the above inequality is non-computable, and therefore, assuming that $e_{\tilde{x}}$ is negligible in comparison with $|\tilde{x}|$, the relative error is defined as follows.

$$\delta(\tilde{x}) \simeq \frac{e_{\tilde{x}}}{|\tilde{x}|} \tag{1.6}$$

1.2.1.9 Example

Suppose $\tilde{x} = 1.41$ and $x = \sqrt{2}$; The relative error of $\tilde{x}$ is as follows

$$\delta(\tilde{x}) = \frac{\sqrt{2} - 1.41}{\sqrt{2}} = 0.002979438000$$

But if we consider $e_{\tilde{x}} = 0.005$, we will have

$$\delta(\tilde{x}) \simeq \frac{e_{\tilde{x}}}{|\tilde{x}|} = 0.003546099291000$$

1.2.1.10 Different Types of Errors Sources

According to what is stated, a list of the error sources is presented in the following.

(A) **The error of the model**. This error includes omissions, ignorance and simplifications to determine the mathematical model of the problem.
(B) **The error of the data**. This error occurs when measuring and estimating problem assumptions.
(C) **The error of the number representation**. Decimal or binary representation of most numbers with finite figures is not possible. Therefore, selecting a finite number of expansion digits of a number causes this error.
(D) **The error of the arithmetic operation**. The result of some operations on two factors has an infinite number of digits, and selecting a finite number of these digits causes this error.
(E) **The error of the method**. Numerical methods are generally iterative and give an approximate of the exact answer. The accuracy of this approximation depends on the type of method and the stopping step.

Among the five error sources mentioned, the error of the model and the error of the data depend on the type of problem, and the people who determine the model of problem in different disciplines are responsible for them. But the next three errors are related to the numerical analysis.

1.2.2 Round of Error and Floating Points Arithmetic

The problem (Sect. 1.2.1.7) is in some way, the same as the theorem (Sect. 1.2.1.1) but is examined from another perspective. Therefore, it is necessary to introduce the definitions and concepts related to floating point. First, we define the floating point.

1.2.2.1 Note

For each computational device, a set of real numbers with finite digits is defined. This set of numbers is called the set of machine numbers and is denoted by the symbol A.

1.2.2.2 Definition

Suppose that $x \notin A$, $rd(x) \in A$ *is* x *approximation if it satisfies the following inequality:*

$$\forall g (g \in A \Rightarrow |x - rd(x)| \leq |x - g|)$$

$rd(x)$ *can be rounded number to* x.

If the computational device has t digits, i.e., the number of Mantis significant digits of machine number is maximum of t digits, then $rd(x)$ will be defined as follows:

$$rd(x) = x(1 + \varepsilon), |\varepsilon| \leq eps$$

where $eps = 5 \times 10^{-t}$ is called machine accuracy.

1.2.2.3 Definition

The scientific representation of a nonzero number such as binary or decimal x is called the floating points representation. In other words, by changing the exponent, the point shifts among the Mantis digits.

The calculations with these numbers are called floating point calculus.

The result of arithmetic operations on machine numbers may or may not be machine numbers, so it cannot be accepted that arithmetic operations in t-digit computers are accurate. For this purpose, we introduce $+ *$, $-^*$, $\times^*$ and $/^*$ as floating points operations, which are as follows for $x, y \in A$:

$$x + *y: = rd(x + y) = (x + y)(1 + \varepsilon_1)$$
$$x - *y: = rd(x - y) = (x - y)(1 + \varepsilon_2)$$
$$x \times *y: = rd(x \times y) = (x \times y)(1 + \varepsilon_3)$$
$$x / * y: = rd(x/y) = (x/y)(1 + \varepsilon_4)$$
$$|\varepsilon_i| \leq eps, i = 1, ...4$$

Therefore, it can be said that floating point arithmetic operations do not satisfy the ordinary arithmetic rules such as having a neutral element, associativity and distributivity.

1.2.2.4 Note

Suppose E is an algebraic term derived from a floating point calculus. In this case, $fl(E)$ is called the value of the term E in the computational device.

The following theorem shows that the error in the truncating method is almost twice the error in the rounding method.

1.2.2.5 Remark

If $fl(x)$ is the floating points approximation of the nonzero number x in a t-digit computer, then:

$$|x - fl(x)| \leq 5 \times |x| \times 10^{-t} \times l$$

where

$$l = \begin{cases} 1 \; rounding \\ 2 \; runcating \end{cases}$$

According to the theorem (Sect. 1.2.2.5), computational devices use the rounding method to store numbers, in which case:

$$\frac{|x - fl(x)|}{|x|} \leq 5 \times 10^{-t} = eps$$

as a result:

$$fl(x) = x(1 + \varepsilon), |\varepsilon| \leq eps \tag{1.7}$$

It can be easily shown that performing floating point operations on numbers will cause the error to propagate and grow, which we will examine in the next section.

Now, we are going to solve some important and applied problems related to the concept of floating point.

1.2.2.6 Problem

Consider the following series:

$$\sum_{i=1}^{n} x_i = x_1 + x_2 + \cdots + x_n$$

Find the floating point approximation of this series.

Solution: Consider:

$$\begin{aligned} s_1 &= x_1 \\ s_r &= fl(s_{r-1} + x_r) \\ &= (s_{r-1} + x_r)(1 + \varepsilon_r), r = 2, ..., n \end{aligned}$$

where $|\varepsilon_r| \leq eps$. In this case:

$$\begin{aligned} s_n &= l(s_{n-1} + x_n)(1 + \varepsilon_n) \\ &= \left[(s_{r-2} + x_{n-1})(1 + \varepsilon_{n-1}) + x_n\right](1 + \varepsilon_n) \\ &\;\vdots \end{aligned}$$

$$= x_1(1+\eta_1) + \cdots + x_n(1+\eta_n)$$

where

$$1+\eta_r = (1+\varepsilon_r)(1+\varepsilon_{r+1})\ldots(1+\varepsilon_n),\ |\varepsilon_r| \le eps,\ r = 2, \ldots, n$$
$$\eta_1 = \eta_2$$

So:

$$(1-eps)^{n-r+1} \le 1+\eta_r \le (1+eps)^{n-r+1}$$

As a result:

$$fl\left(\sum_{i=1}^{n} x_i\right) = \left(\sum_{i=1}^{n} x_i\right)\left(1 + \frac{\sum_{i=1}^{n} x_i\eta_i}{\sum_{i=1}^{n} x_i}\right)$$

1.2.2.7 Problem

Using the t-digit floating point calculus, show:

$$rd(a) = \frac{a}{1+\varepsilon},\ |\varepsilon| \le 5 \times 10^{-t}$$

Solution: We know:

$$rd(a) = a(1+\varepsilon),\ |\varepsilon| \le 5 \times 10^{-t}$$

δ can be defined such that it satisfies the property of ε. For this purpose, it is sufficient to consider δ such that $|\delta| \le 5 \times 10^{-t}$.

We put $\delta = \frac{-\varepsilon}{1+\varepsilon}$, then:

$$rd(a) = a(1+\delta),\ |\delta| \le 5 \times 10^{-t}$$
$$= a\left(1 + \frac{-\varepsilon}{1+\varepsilon}\right),\ |\varepsilon| \le 5 \times 10^{-t}$$
$$= a\left(\frac{1}{1+\varepsilon}\right)$$

1.2.2.8 Problem

Suppose that x is a floating point machine number on a computer with a rounding error unit of ε. Show:

$$fl\left(x^k\right) = x^k(1+\delta)^{k+1},\ |\delta| \le \varepsilon$$

Solution: We prove it by induction on k.

Initial case: Suppose $k = 1$. We have:

$$fl(x) = x(1+\delta)^0 = x$$

Induction hypothesis: Suppose that the statement holds for $k = n > 1$.

Proof of induction hypothesis:

We prove that the statement also holds for $k = n + 1$

$$\begin{aligned} fl\left(x^{n+1}\right) &= fl\left(x^n.x\right) \\ &= fl\left(fl\left(x^n\right).x\right) \\ &= fl\left(x^n\right).x.(1+\delta) \\ &= x^n.(1+\delta)^{n-1}.x.(1+\delta) \\ &= x^{n+1}.(1+\delta)^n \end{aligned}$$

1.2.2.9 Problem

Suppose $|fl(ab) - ab| \le |ab|\beta^{-t+1}$. Calculate the upper bound of the following equation:

$$fl(fl(ab).c) - abc$$

Solution:

$$\begin{aligned} |fl(fl(ab) - abc)| &\le |fl(fl(ab).c) - fl(ab).c| + |fl(ab).c - abc| \\ &\le |fl(ab).c|\beta^{-t+1} + |abc|\beta^{-t+1} \\ &\le (|fl(ab).c - abc| + |abc|)\beta^{-t+1} + |abc|\beta^{-t+1} \\ &\le |abc|\left(1+\beta^{-t+1}\right)\beta^{-t+1} + |abc|\beta^{-t+1} \\ &\le |abc|\left(2+\beta^{-t+1}\right)\beta^{-t+1} \end{aligned}$$

1.2.2.10 Problem

Suppose $S_n = \sum_{i=1}^{n} x_i$ where every x_i is a machine number and suppose that S_n^* is what the computer calculates. In this case, $S_n^* = fl\big(S_{n-1}^* + x_n\big)$. Prove:

$$S_n^* \simeq S_n + S_2\delta_2 + S_3\delta_3 + \cdots + S_n\delta_n, \ |\delta_k| \le eps$$

Also show:

$$\begin{aligned} S_n^* - S_n &\simeq x_1(\delta_2 + \cdots + \delta_n) + x_2(\delta_2 + \cdots + \delta_n) \\ &\quad + x_3(\delta_3 + \cdots + \delta_n) + \cdots + x_n\delta_n \end{aligned}$$

Solution: We put:

$$S_2^* = fl(x_1 + x_2) = (x_1 + x_2)(1 + \delta_2), \ |\delta_2| \le eps$$

And we get by iteration:

$$S_{r+1}^* = fl\big(S_r^* + x_{r+1}\big) = \big(S_r^* + x_{r+1}\big)(1 + \delta_{r+1}), \ |\delta_{r+1}| \le eps, r = 1, \ldots, n-1$$

So, we will have:

$$\begin{aligned} S_2^* - (x_1 + x_2) &= \delta_2(x_1 + x_2) \\ S_3^* - (x_1 + x_2 + x_3) &= fl\big(S_2^* + x_3\big) - (x_1 + x_2 + x_3) \\ &= ((x_1 + x_2)(1 + \delta_2) + x_3)(1 + \delta_3) - (x_1 + x_2 + x_3) \\ &= \delta_2(x_1 + x_2) + \delta_3(x_1 + x_2 + x_3) + \delta_2\delta_3(x_1 + x_2) \\ &\simeq \delta_2(x_1 + x_2) + \delta_3(x_1 + x_2 + x_3) \end{aligned}$$

Finally, it can be concluded using induction that:

$$\begin{aligned} S_n^* &\simeq S_n + \delta_2(x_1 + x_2) + \delta_3(x_1 + x_2 + x_3) + \cdots + \delta_n(x_1 + \cdots + x_n) \\ &= S_n + S_2\delta_2 + \cdots + S_n\delta_n \end{aligned}$$

Also

$$\begin{aligned} S_n^* - S_n &\simeq \delta_2(x_1 + x_2) + \delta_3(x_1 + x_2 + x_3) + \cdots + \delta_n(x_1 + \cdots + x_n) \\ &= x_1(\delta_2 + \cdots + \delta_n) + x_2(\delta_2 + \cdots + \delta_n) \\ &\quad + x_3(\delta_3 + \cdots + \delta_n) + \cdots . + x_n\delta_n \end{aligned}$$

1.2.2.11 Problem

Suppose that $x_0, \ldots, x_n$ are positive machine numbers in a computer with a rounding error of ε. Prove that the relative rounding error in the calculation of $\sum_{i=0}^{n} x_i$, in the ordinary way, is a maximum of $(1+\varepsilon)^n - 1$, which is approximately equal to $n\varepsilon$.

Solution: Suppose that $S_k = x_0 + \cdots + x_k$, and S_k^* is the value that the computer calculates instead of S_k. The iterative formulas for these quantities are:

$$\begin{cases} S_0 = x_0 \\ S_{k+1} = S_k + x_{k+1} \end{cases} \begin{cases} S_0^* = x_0 \\ S_{k+1}^* = fl\left(S_k^* + x_{k+1}\right) \end{cases}$$

For analysis, we define the following values:

$$\rho_k = \frac{S_k^* - S_k}{S_k},\ \delta_k = \frac{S_{k+1}^* - \left(S_k^* + x_{k+1}\right)}{S_k^* + x_{k+1}}$$

Therefore, $|\rho_k|$ is relative errors for approximating the kth partial sum of S_k by the calculated partial sum of S_k^* and $|\delta_k|$ is relative errors in the approximation of $S_k^* + x_{k+1}$ by the quantity $fl\left(S_k^* + x_{k+1}\right)$. Using the relationships that define ρ_k and δ_k, we have:

$$\begin{aligned} \rho_{k+1} &= \frac{\left(S_{k+1}^* - S_{k+1}\right)}{S_{k+1}} \\ &= \frac{\left(S_k^* + x_{k+1}\right)(1+\delta_k) - (S_k + x_{k+1})}{S_{k+1}} \\ &= \frac{\left[S_k(1+\delta_k) + x_{k+1}\right](1+\delta_k) - (S_k + x_{k+1})}{S_{k+1}} \\ &= \delta_k + \rho_k\left(\frac{S_k}{S_{k+1}}\right)(1+\delta_k) \end{aligned}$$

Because $S_k < S_{k+1}$ and $|\delta_k| \le \varepsilon$, we conclude:

$$|\rho_{k+1}| \le \varepsilon + |\rho_k|(1+\varepsilon) = \varepsilon + \theta|\rho_k|$$

where $\theta = 1 + \varepsilon$. So, we have the following consecutive inequalities:

$$\begin{aligned} &|\rho_0| = 0 \\ &|\rho_1| \le \varepsilon \\ &|\rho_2| \le \varepsilon + \theta\varepsilon \\ &|\rho_3| \le \varepsilon + \theta(\varepsilon + \theta\varepsilon) = \varepsilon + \theta\varepsilon + \theta^2\varepsilon \\ &\vdots \end{aligned}$$

The final result is:

$$
\begin{aligned}
|\rho_n| &\le \varepsilon + \theta\varepsilon + \theta^2\varepsilon + \cdots + \theta^{n-1}\varepsilon \\
&= \varepsilon\left(1 + \theta + \theta^2 + \cdots + \theta^{n-1}\right) \\
&= \varepsilon.\frac{\theta^n - 1}{\theta - 1} \\
&= \varepsilon.\frac{(1+\varepsilon)^n - 1}{\varepsilon} \\
&= (1+\varepsilon)^n - 1
\end{aligned}
$$

According to the binomial theorem, we have:

$$
(1+\varepsilon)^n - 1 = 1 + \binom{n}{1}\varepsilon + \binom{n}{2}\varepsilon^2 + \ldots - 1 \approx n\varepsilon
$$

1.2.2.12 Problem

Suppose that $f(x) = x^{2^n}$ and $x = x_0$ is not machine number. So, on the computer, it will have a modified value of $x = x_0(1+\delta)$. Estimate how much $f(x_0)$ will change?

Solution: It is obvious that the value of function $f(x)$ at point x_0 is calculated using n squares by performing a sequence of calculations as follows:

$$
x_1 := x_0^2,\, x_2 := x_1^2,\, \ldots,\, x_n := x_{n-1}^2
$$

which will eventually be equal to $f(x_0) = x_n$. According to the floating point calculus, we will have:

$$
\widehat{x}_1 = x_0^2(1+\delta_1),\, \widehat{x}_2 = x_1^2(1+\delta_2),\, \ldots,\, \widehat{x}_n = x_{n-1}^2(1+\delta_n)
$$

where for $i = 1, \ldots, n$, $|\delta_i| \le eps$. The nested substitution of relationships can result in:

$$
\widehat{x}_n = x_0^{2^n}(1+\delta_n)^{2^{n-1}} \ldots (1+\delta_{n-1})^2(1+\delta_n)
$$

Now, we suppose that $\delta = \max\limits_{1\le i\le n} \delta_i$, so:

$$
\widehat{x}_n \approx x_0^{2^n}(1+\delta_n)^{2^n-1}
$$

We define η such that $|\eta| \le eps$ and

$$(1+\delta)^r \approx (1+\eta)^{r+1}$$

So, we conclude:

$$\hat{x}_n \approx x_0^{2^n}(1+\eta)^{2^n} = f(x_0(1+\eta))$$

1.2.3 Algorithm Error Propagation

In this section, we examine the propagation of rounding error in an algorithm. As mentioned earlier,

$$\phi = \varphi^{(r)} \circ \ldots \circ \varphi^{(0)}$$

If we define $X^{(0)} = X$, then, Y is obtained as follows:

$$X = X^{(0)} \to \varphi^{(0)}\big(X^{(0)}\big) \to \ldots \to \varphi^{(r)}\big(X^{(r)}\big) = X^{(r+1)} = Y \tag{1.8}$$

Suppose that we have the map $\psi^{(i)}$ (also called the remaining map) as follows:

$$\psi^{(i)} = \varphi^{(r)} \circ \ldots \circ \varphi^{(0)},\ \psi^{(i)} : D_i \to \mathbb{R}^m,\ i = 0, \ldots, r$$

where $\psi^{(0)} \equiv \phi$, which means that none of algorithm steps have been performed. So, for $i = 0, \ldots, r$, we have:

$$D\psi^{(i)}\big(X^{(i)}\big) = D\varphi^{(r)}\big(X^{(r)}\big).D\varphi^{(r-1)}\big(X^{(r-1)}\big)\ldots D\varphi^{(i)}\big(X^{(i)}\big) \tag{1.9}$$

where $D\psi^{(i)}$ is a Jacobian matrix of map $\psi^{(i)}$. Therefore, the propagation of rounding error in the steps of performing an algorithm can be defined as follows:

$$\begin{aligned}
\Delta\tilde{X}^{(1)} &\simeq D\varphi^{(0)}(X).\Delta\tilde{X} + \alpha_1 \\
\Delta\tilde{X}^{(2)} &\simeq D\varphi^{(1)}\big(X^{(1)}\big)\Big(D\varphi^{(0)}(X).\Delta\tilde{X} + \alpha_1\Big) + \alpha_2 \\
&\ \ \vdots \\
\Delta\tilde{X}^{(r+1)} &= \Delta\tilde{Y} \\
&\simeq D\varphi^{(r)}\big(X^{(r)}\big)\ldots D\varphi^{(0)}(X).\Delta\tilde{X} + D\varphi^{(r)}\big(X^{(r)}\big)\ldots D\varphi^{(1)}\big(X^{(1)}\big).\alpha_1 \\
&\quad + \cdots + D\varphi^{(r)}\big(X^{(r)}\big).\alpha_r + \alpha_{r+1}
\end{aligned}$$

Now if for $i = 0, \ldots, r$ we define $\alpha_{i+1} := E_{i+1}.X^{(i+1)}$, where

$$E_{i+1} := \begin{bmatrix} \varepsilon_1 & \cdots & 0 \\ \vdots & \ddots & \vdots \\ 0 & \ldots & \varepsilon_{n_i+1} \end{bmatrix}$$

is the error diagonal matrix, we will have:

$$\begin{aligned} \Delta\tilde{Y} &= D\phi(X).\Delta\tilde{X} + D\psi^{(1)}\big(X^{(1)}\big).E_1.X^{(1)} \\ &\quad + \cdots + D\psi^{(r)}\big(X^{(r)}\big).E_r.X^{(r)} + E_{r+1}.Y \end{aligned} \tag{1.10}$$

Now we suppose that no step of the algorithm is executed, so in Eq. (1.10), we will have only $D\phi(X).\Delta\tilde{X}$ and $E_{r+1}.Y$. So $\Delta\tilde{Y}$ will be as follows

$$\begin{aligned} \Delta^{(0)}\tilde{Y} &= D\phi(X).\Delta\tilde{X} + +E_{r+1}.Y \\ &= (|D\phi(X)|.|X| + |Y|)eps \end{aligned}$$

which is called the inherent error of the algorithm. It is clear that the inherent error depends on the input vector of X and the output vector of Y will appear in any case.

If for every i, the rounding error propagations (E_i's) on the total error are such that:

$$\left|D\psi^{(i)}\big(X^{(i)}\big).E_i.X^{(i)}\right| \simeq \Delta^{(0)}\tilde{Y}$$

Then, we say the rounding errors are harmless. If the rounding errors are harmless at all steps, the algorithm is well-established or numerically stable.

1.2.3.1 Problem

The following function is calculated for $0 \le \theta \le 2\pi$ and $0 < k_c \le 1$:

$$f(\theta, k_c) := \frac{1}{\sqrt{\cos^2\theta + k_c^2 \sin^2\theta}}$$

If we apply the following method:

$$\begin{aligned} k^2 &:= 1 - k_c^2 \\ f(\theta, k_c) &:= \frac{1}{\sqrt{1 - +k^2 \sin^2\theta}} \end{aligned}$$

In this case, we do not need to calculate $\cos\theta$ and the method is faster. Compare both methods for numerical stability, as follows.

Solution: Algorithm 1:

$$\phi(a, b) = \cos^2 a + b\sin^2 a$$

If $x = x^{(0)} = \begin{bmatrix} a \\ b \end{bmatrix}$, then:

$$\varphi^{(0)}\left(x^{(0)}\right) = \begin{bmatrix} \cos a \\ \sin a \\ b \end{bmatrix} = x^{(1)}$$

$$\varphi^{(1)}\left(x^{(1)}\right) = \begin{bmatrix} \cos^2 a \\ \sin^2 a \\ b \end{bmatrix} = x^{(2)}$$

$$\varphi^{(2)}\left(x^{(2)}\right) = \begin{bmatrix} \cos^2 a \\ b\sin^2 a \end{bmatrix} = x^{(3)}$$

$$\varphi^{(3)}\left(x^{(3)}\right) = \cos^2 a + b\sin^2 a = x^{(4)} = \phi(a, b)$$

So, we have:

$$\psi^{(1)}(u, v, w) = \varphi^{(3)} \circ \varphi^{(2)} \circ \varphi^{(1)}(u, v, w) = u^2 + wv^2$$

Then:

$$D\psi^{(1)}(u, v, w) = \left(2u, 2wv, v^2\right)$$

By substituting $x^{(1)}$ we have:

$$D\psi^{(1)}\left(x^{(1)}\right) = \left(2\cos a, 2b\sin a, \sin^2 a\right)$$

and also:

$$\psi^{(2)}(r, s, t) = \varphi^{(3)} \circ \varphi^{(2)}(r, s, t) = r + st$$

So:

$$D\psi^{(2)}(r, s, t) = (1, s, t)$$

By considering $x^{(2)}$, we have:

$$D\psi^{(2)}\left(x^{(2)}\right) = \left(1, b, \sin^2 a\right)$$

and also:

$$\psi^{(3)}(m, n) = m + n$$

So:

$$D\psi^{(3)}(m, n) = (1, 1)$$

By substituting $x^{(3)}$, you can write:

$$D\psi^{(3)}\left(x^{(3)}\right) = (1, 1)$$

and finally:

$$D\phi(x) = D\phi(a, b) = \left(-2 \sin a \cos a + 2b \sin a \cos a, \sin^2 a\right)$$

and according to the above relations, it can be written:

$$\alpha_1 = \begin{bmatrix} \varepsilon_1 \cos a \\ \varepsilon_2 \sin a \\ 0 \end{bmatrix}, \alpha_2 = \begin{bmatrix} \varepsilon_3 \cos^2 a \\ \varepsilon_4 \sin^2 a \\ 0 \end{bmatrix}$$

$$\alpha_3 = \begin{bmatrix} 0 \\ \varepsilon_5 b \sin^2 a \end{bmatrix}, \alpha_4 = \varepsilon_6\left(\cos^2 a + b \sin^2 a\right)$$

Since:

$$\Delta\tilde{x} = |\tilde{x} - x| = |fl(x) - x| = |x(1 + \varepsilon) - x| = |x|\varepsilon, |\varepsilon| \le eps \tag{1.11}$$

So finally we will have:

$$\begin{aligned}
\Delta\tilde{y} &= D\phi(x)\Delta\tilde{x} + D\psi^{(1)}\left(x^{(1)}\right)\alpha_1 + D\psi^{(2)}\left(x^{(2)}\right)\alpha_2 + D\psi^{(3)}\left(x^{(3)}\right)\alpha_3 + \alpha_4 \\
&= 2(b - 1) \sin a \cos a.\Delta a + \sin^2 a.\Delta b + 2\varepsilon_1 \cos^2 a \\
&\quad + 2\varepsilon_2 \sin^2 a + \varepsilon_3 \cos^2 a + b\varepsilon_4 \sin^2 a \\
&\quad + b\varepsilon_5 \sin^2 a + \varepsilon_6 \cos^2 a + b\varepsilon_6 \sin^2 a \\
&\le 2(b - 1) \sin a \cos a a|\varepsilon| + \sin^2 a b|\varepsilon| \\
&\quad + 2\cos^2 a|\varepsilon_1| + 2b \sin^2 a|\varepsilon_2| + \cos^2 a|\varepsilon_3| \\
&\quad + 2b \sin^2 a|\varepsilon_4| + 2b \sin^2 a|\varepsilon_5| + \cos^2 a|\varepsilon_6| + 2b \sin^2 a|\varepsilon_6| \\
&\le 3eps + \left(4\cos^2 a + 5b \sin^2 a\right).eps
\end{aligned}$$

But about Algorithm 2:

If $x = x^{(0)} = \begin{bmatrix} a \\ b \end{bmatrix}$, then:

$$\varphi^{(0)}\left(x^{(0)}\right) = \begin{bmatrix} \sin a \\ 1 - b \end{bmatrix} = x^{(1)}$$

$$\varphi^{(1)}\left(x^{(1)}\right) = \begin{bmatrix} \sin^2 a \\ 1 - b \end{bmatrix} = x^{(2)}$$

$$\varphi^{(2)}\left(x^{(2)}\right) = (1 - b)\sin^2 a = x^{(3)}$$

$$\varphi^{(3)}\left(x^{(3)}\right) = 1 - (1 - b)\sin^2 a = x^{(4)} = \phi(a, b)$$

So:

$$\psi^{(1)}(m, n) = \varphi^{(3)} \circ \varphi^{(2)} \circ \varphi^{(1)}(m, n) = 1 - m^2 n$$

Then:

$$D\psi^{(1)}(m, n) = \left(-2mn, -m^2\right)$$

By substituting $x^{(1)}$ we have:

$$D\psi^{(1)}\left(x^{(1)}\right) = \left(-2(1 - b)\sin a, -\sin^2 a\right)$$

and also:

$$\psi^{(2)}(r, s) = \varphi^{(3)} \circ \varphi^{(2)}(r, s) = 1 - rs$$

So:

$$D\psi^{(2)}(r, s) = (-s, -r)$$

By substituting $x^{(2)}$, we have:

$$D\psi^{(2)}\left(x^{(2)}\right) = \left(-(1 - b), -\sin^2 a\right)$$

and also:

$$\psi^{(3)}(t) = 1 - t$$

So:

$$D\psi^{(3)}(t) = -1$$

By substituting $x^{(3)}$, you can write:

$$D\psi^{(3)}\left(x^{(3)}\right) = -1$$

and finally:

$$D\phi(x) = D\phi(a,b) = \left(-2(1-b)\sin a\cos a, \sin^2 a\right)$$

and according to the above relations, it can be written:

$$\alpha_1 = \begin{bmatrix} \varepsilon_1 \sin a \\ \varepsilon_2(1-b) \end{bmatrix}, \alpha_2 = \begin{bmatrix} \varepsilon_3 \sin^2 a \\ 0 \end{bmatrix}$$
$$\alpha_3 = \varepsilon_4(1-b)\sin^2 a, \alpha_4 = \varepsilon_5\left(1-(1-b)\sin^2 a\right)$$

Given the relation (1.11), we will finally have:

$$\begin{aligned}\Delta\tilde{y} &= D\phi(x)\Delta\tilde{x} + D\psi^{(1)}\left(x^{(1)}\right)\alpha_1 + D\psi^{(2)}\left(x^{(2)}\right)\alpha_2 + D\psi^{(3)}\left(x^{(3)}\right)\alpha_3 + \alpha_4 \\ &= 2(b-1)\sin a\cos a.\Delta a + \sin^2 a.\Delta b \\ &\quad + 2(b-1)\varepsilon_1\sin^2 a + (b-1)\varepsilon_2\sin^2 a + (b-1)\varepsilon_3\sin^2 a \\ &\quad + (b-1)\varepsilon_4\sin^2 a + \varepsilon_5\left(1-(1-b)\sin^2 a\right) \\ &\le 2(b-1)\sin a\cos a a|\varepsilon| + \sin^2 a b|\varepsilon| \\ &\quad + \left(2(1-b)\sin^2 a + (1-b)\sin^2 a + (1-b)\sin^2 a + \ldots\right).eps \\ &= 3eps + \left(6(1-b)\sin^2 a + 1\right).eps\end{aligned}$$

For the Algorithm 2 to be numerically better than the Algorithm 1, we must have:

$$\begin{aligned}&6(1-b)\sin^2 a + 1 \le 4\cos^2 a + 5b\sin^2 a \\ &\Rightarrow 6(1-b)\sin^2 a + 1 - 4\cos^2 a - 5b\sin^2 a \le 0 \\ &\Rightarrow \sin^2 a(6-6b-5b) - 4\cos^2 a + 1 \le 0\end{aligned}$$

Given that $0 \le a \le \frac{\pi}{2}$ 1, so $\sin a \le 1$ and $\cos a \le 1$. So:

$$6 - 11b \le 3 \Rightarrow b \ge \frac{3}{11}$$

So, if $\frac{3}{11} < b < 1$, the Algorithm 2 is better than the Algorithm 1, otherwise, the Algorithm 1 would be better, that is, if $0 < b < \frac{3}{11}$, the Algorithm 1 is better. If $b = \frac{3}{11}$ and $0 \le a \le \frac{\pi}{2}$, the Algorithm 2 still works better than the Algorithm 1.

1.2.3.2 Scientific Representation of Numbers

Suppose that x is a nonzero number. It is obvious that x can always be written as

$$x = p.10^q, 1 \leq |p| < 10$$

where y is an integer.

In this case, we can state that x is scientifically represented. In this display, p is called the mantissa and q is called the exponent of the number x.

1.2.3.3 Definition

If p is a decimal number and $1 \leq |p| < 10$, then the significant figures of p are the nonzero figures of p, the zeros between these figures and the zeros in front of the number to indicate the accuracy. The significant figures of the nonzero number x are same as the significant figures of the mantissa of x.

1.2.3.4 Example

(A) If $x = 301.57$, then $x = 3.0157 \times 10^2$ and the number of significant figures of x is 5.
(B) If $x = 0.000497$, then $x = 4.97 \times 10^{-4}$; and x has three significant figures.
(C) If $x = 2000m$, then $x = 2.000 \times 10^3 m$; and x has four significant figures.

And especially since always $\frac{n+1}{n} > 1$, none of the above numbers is equal to the sequence limit. In numerical analysis, we are always faced with such a situation, that is, a sequence of numbers is created that converges to the answer of the considered problem under certain conditions. Thus, we should generally define the error of an arbitrary approximation of a number. For this purpose, we assume that x is a number (accurate) and $\tilde{x}$ is an approximation of that.

1.3 Interpolation

Considering the many applications of mathematical functions in different methods, it is important to have a function rule. Sometimes, we may only have some data from a function, and if we want to know the behavior of the function, we must have a graph of the function. For example, suppose that a surveying team takes coordinates from a long path with surveying cameras. For instance, this path could be an airport runway or a tunnel between cities or a subway line between city stations. The coordinates obtained from this survey are our data or assumptions. If we know the behavior of these coordinates, that is, if we have a curve connecting these points, then we will have the same airport runway or tunnel or subway line. We know that the aircraft applies a strong impulse to the runway when it lands on the runway and then starts to stop at a high speed of about 250 km per hour. Obviously, if the runway has small bumps, it will shorten the life of the aircraft. Therefore, it is necessary for the runway to be a very smooth path with the least curvature. For this purpose, the function that approximates the behavior of these coordinates must be a smooth, low-curvature function. This can be done using a very powerful interpolation function such as spline. Sometimes, as we consider the approximate location of points or coordinates, the drawn curve may not pass through some points, in which case, we will have another form of approximation called the curve fitting. In both cases, for better approximation, we select an interpolation function that behaves similar to the approximate behavior of the points and their distribution. Recently, the topic of interpolation, specially, the splines, has been introduced in medical sciences. For example, it is used in very fine devices that take films or photographs from the inside of the intestine and stomach. Other applications of this type of interpolation function are in various engineering sciences, of which everyone is aware.

A general but brief introduction to interpolation functions is given below.

Assume that the values of the function f are $f_0, ..., f_n$, respectively, for the mutual distinct points of $x_0, ..., x_n$. We call such a function a tabular function. Estimating the value of $f(x)$ when $x \in [x_0, x_n]$ and $x \neq x_i, i = 0, ..., n$ is called interpolation and estimating the value of $f(x)$ when $x \in [x_0, x_n]$ is called extrapolation.

Estimating the value of a function for a point like x that is not in the table but between table points is also an interpolation.

Suppose that ϕ is a family of single-variable functions of x with $n + 1$ parameters of $a_0, ..., a_n$. The problem of interpolation ϕ is to determine the parameters $a_0, ..., a_n$ in $\phi(x; a_0, ..., a_n)$ so that for pairs $(x_i, f_i), i = 0, ..., n$ we have:

$$\phi(x_i; a_0, ..., a_n) = f_i, i = 0, ..., n \tag{1.12}$$

where the mutual distinct points of $(x_i, f_i), i = 0, ..., n$ can be real or complex. Equation (1.12) is called the interpolation problem and points (x_i, f_i) are called the interpolation points.

Obviously, the goal is to determine the $n + 1$ unknown parameters of $a_0, ..., a_n$ using the set of equations (1.12). If ϕ is linearly dependent on parameters $a_0, ..., a_n$, then the interpolation problem (1.12) is called linear, i.e.,

$$\phi(x; a_0, ..., a_n) = a_0\phi_0(x) + \cdots + a_n\phi_n(x) \tag{1.13}$$

Otherwise, call it nonlinear. For example:

Linear:

(1) If $\phi_j(x) = x^j, j = 0, ..., n$, then $\phi(x; a_0, ..., a_n)$ defines an interpolator as a polynomial of at most degree n.

(2) If $\phi_j(x) = e^{ijx}$ or $\phi_j(x) = \cos jx$ or $\phi_j(x) = \sin jx, j = 0, ..., n$, then $\phi(x; a_0, ..., a_n)$ is a triangular interpolator.

(3) If $\phi \in c^2[x_0, x_n]$, then $\phi|_{[x_j, x_{j+1}]} \in \Pi_3$ is a cubic spline interpolator.

Nonlinear:

(4) If $\phi(x; a_0, ..., a_\mu, b_0, ..., b_\nu) = \frac{a_0+a_1x+\cdots+a_\mu x^\mu}{b_0+b_1x+\cdots+b_\nu x^\nu}$, ϕ is a fractional interpolator.

(5) If $\phi(x; a_0, ..., a_\mu, \lambda_0, ..., \lambda_\mu) = a_0 e^{\lambda_0 x} + \cdots + a_\mu e^{\lambda_\mu x}$, then ϕ is an exponential interpolator.

It is worth noting that for each of interpolations that have been proposed, the introduced interpolation of ϕ is a linear combination of a set of functions of $\{\phi_0, ..., \phi_n\}$, that is, they are generated by its members. These types of functions are defined so that they form an independent linear set. Therefore, it can be claimed that all interpolator functions form a vector space that can have different bases, and each of the interpolators, according to the structure of the basic members, belongs to their own space. Obviously, the representation of each interpolation function by the basic members in the corresponding space is unique, so generally, in Eq. (1.13), the linear combination coefficients of a_i, $i = 0, ..., n$ are obtained uniquely. Therefore, it can be concluded that any kind of interpolator is unique in its own space. If we want to have a kind of classification in the order of the appearance of the interpolation functions, we can say:

(A) Non-recursive interpolations
Example: Lagrange, trigonometric, fractional and exponential interpolators.
(B) Recursive interpolations
Example: Neville, Aitken and Newton interpolators.
(C) Interpolation with multicriteria functions
Example: Hermit, spline and B-spline interpolators.

We will now concisely discuss each of these interpolations.

1.3.1 Lagrange Interpolation

This type of interpolation is a special case of type (A) interpolation.

Polynomial function of

$$p(x) = \sum_{j=0}^{n} L_j(x) f_j \tag{1.14}$$

is a Lagrange interpolator polynomial if it holds in the interpolation condition of $p(x_i) = f_i$, $i = 0, 1, ..., n$, where

$$L_j(x) = \prod_{i=0, j \neq j}^{n} \frac{x - x_i}{x_j - x_i}, \ j = 0, ..., n \tag{1.15}$$

are Lagrange polynomials of degree n with condition of $L_j(x_k) = \delta_{jk}$, $0 \leq j, k \leq n$. The calculations in the Lagrange method, even if n is not too large, are huge and time consuming, and automation of the operation is not easy. Also, the degree of interpolation polynomials is determined after the calculations are completed. By adding a point to the table, all operations must be repeated, and since interpolation polynomials are not calculated gradually, this method must be used with caution. Now, we consider some properties of this interpolation using some problems.

1.3.1.1 Problem

Prove that

$$\sum_{i=0}^{n} L_i(x) = 1$$

Solution: Suppose that $f(x) = 1$, and the interpolation polynomials $p(x)$ for points of $x_0, ..., x_n$ are such that $p \in \Pi_n$ and for $i = 0, ..., n$, $p(x_i) = 1$. On the other hand, according to Eq. (1.14) we have:

$$p(x) = \sum_{i=0}^{n} f_i L_i(x) = \sum_{i=0}^{n} L_i(x)$$

By contradiction, suppose that $p(x) \equiv 1$ is not hold, then equation $p(x) - 1 = 0$ has the $n + 1$ roots of $x_0, ..., x_n$, while according to the above discussion, $p \in \Pi_n$, so $p(x) - 1 = 0$ is a polynomial of at most degree n and cannot have $n + 1$ roots. As a result, the assumed statement is false and $p(x) \equiv 1$ and $\sum_{i=0}^{n} L_i(x) = 1$.

1.3.1.2 Problem

If $w(x) = \prod_{j=0}^{n} (x - x_j)$, show

$$L_i(x) = \frac{w(x)}{(x - x_i)w'(x_i)}$$

Solution: According to the definition of the derivative of the function in the point x_i, we have:

$$w'(x_i) = \lim_{x \to x_i} \frac{w(x) - w(x_i)}{x - x_i} = \lim_{x \to x_i} \frac{w(x)}{x - x_i} = \prod_{j=0, j \neq i}^{n} (x_i - x_j)$$

According to Eq. (1.15):

$$L_i(x) = \prod_{j=0, j\neq i}^{n} \left(\frac{x - x_j}{x_i - x_j} \right)$$

So, we can write:

$$L_i(x) = \frac{\prod_{j=0, j\neq i}^{n} (x - x_j)}{\prod_{j=0, j\neq i}^{n} (x_i - x_j)} = \frac{\prod_{j=0}^{n} (x - x_j)}{(x - x_j)\prod_{j=0, j\neq i}^{n} (x_i - x_j)} = \frac{w(x)}{(x - x_j)w'(x_i)}$$

1.3.1.3 Problem

Suppose $h(x) = L_i(x) + L_{i+1}(x)$ for $0 \leq i < n$. Prove that for every x of the interval $[x_i, x_{i+1}]$ we have:

$$h(x) = L_i(x) + L_{i+1}(x) \geq 1$$

Solution: By contradiction, assume that:

$$\exists x (x \in (x_i, x_{i+1}) \Rightarrow L_i(x) + L_{i+1}(x) < 1, 0 \leq i \leq n - 1)$$

Consider the following partition:

$$x_0 < x_1 < \cdots < x_{i-1} < x_i < x < x_{i+1} < \cdots < x_{n-1} < x_n$$

According to the Rolle's theorem, for each of the subintervals of $(x_i, x_{i+1}), i = 0, 1, ..., i-1, i+2, ..., n-1, h'(x) = 0$ has one root, and since the number of these subintervals is $n - 3$, then $h'(x) = 0$ has $n - 3$ roots in the mentioned subintervals. Now according to the average value theorem:

$$\exists \beta_1 \left(\beta_1 \in (x_{i-1}, x_i) \& h'(\beta_1) = \frac{h(x_i) - h(x_{i-1})}{x_i - x_{i-1}} > 0 \right)$$

$$\exists \beta_2 \left(\beta_2 \in (x_i, x) \& h'(\beta_2) = \frac{h(x) - h(x_i)}{x - x_i} < 0 \right)$$

$$\exists \beta_3 \left(\beta_3 \in (x, x_{i+1}) \& h'(\beta_3) = \frac{h(x_{i+1}) - h(x)}{x_{i+1} - x} > 0 \right)$$

$$\exists \beta_4 \left(\beta_4 \in (x_{i+1}, x_{i+2}) \& h'(\beta_4) = \frac{h(x_{i+2}) - h(x_{i+1})}{x_{i+2} - x_{i+1}} < 0 \right)$$

According to the median value theorem, it can be written:

$$\exists\theta_1\big(\theta_1 \in (\beta_1, \beta_2)\&h'(\theta_1) = 0\big)$$
$$\exists\theta_2\big(\theta_2 \in (\beta_2, \beta_3)\&h'(\theta_2) = 0\big)$$
$$\exists\theta_3\big(\theta_3 \in (\beta_3, \beta_4)\&h'(\theta_3) = 0\big)$$

Therefore, $h'(x) = 0$ has three other roots. As a result, totally, it will have n roots. But since $h(x)$ is of at most degree n, then $h'(x)$ is of at most degree $n-1$ and cannot have n roots. Therefore, the assumption is invalid and $h(x) \geq 1$.

1.3.1.4 Problem

Prove that Lagrange polynomials are linearly independent.

Solution: Suppose that for every x:

$$a_0 L_0\big(x_j\big) + a_1 L_1\big(x_j\big) + \cdots + a_j L_j\big(x_j\big) + \cdots + a_n L_n\big(x_j\big) = 0$$

We should prove $a_0 = a_1 = \cdots = a_n = 0$.
By substituting x_j in the above relation we have:

$$a_0 L_0(x) + a_1 L_1(x) + \cdots + a_n L_n(x) = 0$$

Since the Lagrange polynomials satisfy the Kronecker delta condition, then the left hand of the above equation is equal to a_j. As a result:

$$a_j = 0,\ j = 0, \ldots, n$$

1.3.1.5 Problem

Prove that the following set is linearly independent.

$$\{1, (x - x_0), (x - x_0)(x - x_1), \ldots, (x - x_0)\ldots(x - x_{n-1})\}$$

Solution: Suppose that for every x,
We prove that $a_0 = a_1 = \cdots = a_n = 0$. By contradiction, we suppose that i is the first index for which $a_i \neq 0$. So, for every x we have:

$$a_0 + a_1(x - x_0) + \cdots + a_i(x - x_0)\ldots(x - x_{i-1})$$
$$+ \cdots + a_n(x - x_0)\ldots(x - x_{n-1}) = 0$$

Now if we put substitute x by x_i, we get:

$$a_i(x_i - x_0)\ldots(x_i - x_{i-1}) = 0$$

Because the points are mutually distinctive, then $a_{i=0}$ which is contrary to the assumption. As a result, the assumed statement is invalid and $a_0 = a_1 = \cdots = a_n = 0$.

1.3.2 *Newton's Divided Difference Interpolation*

This type of interpolation is type (A) interpolation based on the Neville recursive method. The polynomial

$$p(x) = f_0 + (x - x_0)f[x_0, x_1] + \cdots + (x - x_0)\ldots(x - x_{n-1})f[x_0, \ldots, x_n]$$

is the Newton interpolation polynomial where

$$f[x_0, \ldots, x_n] = \frac{f[x_1, \ldots, x_n] - f\left[x_0, \ldots, x_{n-1}\right]}{x_n - x_0}$$

are nth order divided differences between points $x_0, \ldots, x_n$.

It should be noted that the Newton's divided differences are invariant under permutations of the indices, and the previous calculations can still be used by adding a point. So, the coefficients of this interpolation polynomial are stable.

1.3.2.1 Problem

Suppose that $f(x)$ is a polynomial of degree n, then:

$$\forall k(k > n \Rightarrow f[x_0, \ldots, x_k] = 0)$$

Solution: We assume that $f(x) = x^n$ without losing generality. It is clear that $f(x)$ passes through interpolation points, so it can be considered as an interpolation polynomial. On the other hand, if we want to write the divided differences interpolation polynomial formula, due to the uniqueness of the interpolation polynomial, we have:

$$\begin{aligned} x^n = {} & f_0 + (x - x_0)f[x_0, x_1] + \cdots + (x - x_0)\ldots(x - x_{n-1})f[x_0, \ldots, x_n] \\ & + \cdots + (x - x_0)\ldots(x - x_{k-1})f[x_0, \ldots, x_k] \end{aligned}$$

Therefore, the assertion is hold.

1.3.2.2 Problem

Prove that if $x_i = x_0 + ih$ for $i = 0, \dots, n$, then:

$$\Delta^n f(x_0) = n!h^n f[x_0, \dots, x_n]$$

Solution: We prove the above relation by induction on n.

Initial case: Suppose $n = 1$.

$$\Delta f(x_0) = f(x_1) - f(x_0) = (x_1 - x_0)\frac{f(x_1) - f(x_0)}{x_1 - x_0} = hf[x_0, x_1]$$

Induction hypothesis: We suppose that the above relation holds for $n \geq 1$.

Induction step: We prove that it holds for $n + 1$ as well.

$$\begin{aligned}
\Delta^{n+1} f(x_0) &= \Delta^n f(x_1) - \Delta^n f(x_0) \\
&= n!h^n f\left[x_1, \dots, x_{n+1}\right] - n!h^n f[x_0, \dots, x_n] \\
&= n!h^n f(x_{n+1} - x_0)\frac{f\left[x_1, \dots, x_{n+1}\right] - f[x_0, \dots, x_n]}{x_{n+1} - x_0} \\
&= (n+1)!h^{n+1} f\left[x_0, \dots, x_{n+1}\right]
\end{aligned}$$

1.3.2.3 Problem

Prove that:

$$f[x_0, \dots, x_n] = \sum_{i=0}^{n} \frac{f(x_i)}{\prod_{j=0, j\neq i}^{n}(x_i - x_j)}$$

Solution: According to the Lagrange interpolation formula, we have:

$$\begin{aligned}
p(x) &= \sum_{i=0}^{n} f_i L_i(x) \\
&= \sum_{i=0}^{n} f_i \prod_{j=0, j\neq i}^{n} \frac{x - x_j}{x_i - x_j} \\
&= \sum_{i=0}^{n} \prod_{j=0, j\neq i}^{n} (x - x_j)\frac{f_i}{\prod_{j=0, j\neq i}^{n}(x_i - x_j)}
\end{aligned}$$

On the other hand, according to Newton's interpolation formula, we have:

$$p(x) = f(x_0) + (x - x_0)f[x_0, x_1] + \cdots + (x - x_0)...(x - x_{n-1})f[x_0, ..., x_n]$$

Because the interpolation polynomial is unique, the coefficients of the leading terms must be equal, then:

$$f[x_0, ..., x_n] = \sum_{i=0}^{n} \frac{f(x_i)}{\prod_{j=0, j\neq i}^{n} (x_i - x_j)}$$

1.3.2.4 Problem

Suppose that $x_j = x_0 + jh$ for $j = 0, ..., n$. In this case, prove that:

$$\Delta^n f(x_0) = \sum_{i=0}^{n} (-1)^{n-i} C_n^i f(x_i)$$

Solution: Given that $x_j = x_0 + jh$, $j = 0, ..., n$, then:

$$\begin{aligned}
\prod_{j=0, j\neq i}^{n} (x_i - x_j) &= \prod_{j=0, j\neq i}^{n} (i - j)h \\
&= \prod_{j=0}^{i-1} (i - j)h \cdot \prod_{l=i+1}^{n} (i - l)h \\
&= (h \cdot 2h \cdots ih)((-h) \cdot (-2h) \cdots (-(n - i)h)) \\
&= (-1)^{n-1} i!(n - i)!h^n
\end{aligned}$$

According to the problem (Sect. 1.3.2.3), we have:

$$\begin{aligned}
f[x_0, ..., x_n] &= \sum_{i=0}^{n} \frac{f(x_i)}{\prod_{j=0, j\neq i}^{n} (x_i - x_j)} \\
&= \sum_{i=0}^{n} \frac{f(x_i)}{(-1)^{n-i} i!(n - i)!h^n} \\
&= \frac{1}{n!h^n} \sum_{i=0}^{n} (-1)^{n-i} C_n^i f(x_i)
\end{aligned}$$

So, according to the problem (Sect. 1.3.2.3), we will have:

$$\Delta^n f(x_0) = n!h^n f[x_0, ..., x_n]$$

$$= \sum_{i=0}^{n} (-1)^{n-i} C_n^i f(x_i)$$

1.3.2.5 Problem

Assume that the nth derivatives of the function $f(x)$ in $I[x_0, ..., x_n]$ are continuously exist. If $x_0, ..., x_n$ are distinct points, then:

$$f[x_0, ..., x_n] = \int_0^1 dt_1 \int_0^{t_1} dt_2 \cdots \int_0^{t_{n-1}} dt_n$$
$$\times f^{(n)}(t_n(x_n - x_{n-1}) + \cdots + t_1(x_1 - x_0) + x_0)$$

where are $n \geq 1$ and $t_0 = 1$.

Solution: By induction on n, we prove the above relation.

Initial case: If $n = 1$, we must prove:

$$f[x_0, x_1] = \int_0^1 dt_1 \times f'(t_1(x_1 - x_0) + x_0)$$

For this purpose, we introduce a new integration variable named ξ for which

$$\xi = t_1(x_1 - x_0) + x_0, dt_1 = \frac{d\xi}{x_1 - x_0}$$

The integration bounds change as follows.

$$t_1 = 0 \Rightarrow \xi = x_0$$
$$t_1 = 1 \Rightarrow \xi = x_1$$

So, we will have

$$\int_0^1 dt_1 \times f'(t_1(x_1 - x_0) + x_0) = \frac{1}{x_1 - x_0} \int_{x_0}^{x_1} d\xi \times f'(\xi)$$
$$= \frac{f(x_1) - f(x_0)}{x_1 - x_0}$$
$$= f[x_0, x_1]$$

Inductive hypothesis: Suppose that the relation holds for $n - 1$, i.e.,

$$f[x_0, \ldots, x_{n-1}] = \int_0^1 dt_1 \int_0^{t_1} dt_2 \cdots \int_0^{t_{n-2}} dt_{n-1} \times f^{(n-1)}(t_{n-1}(x_{n-1} - x_{n-2}) + \cdots + t_1(x_1 - x_0) + x_0)$$

Induction step: We prove that the mentioned relation is hold for n.
To prove, we place the following relation.

$$\xi = t_n(x_n - x_{n-1}) + \cdots + t_1(x_1 - x_0) + x_0,\ dt_n = \frac{d\xi}{x_n - x_{n-1}}$$

Therefore, the integration bounds will be as follows:

$$\begin{aligned}
t_n &= 0 \Rightarrow \xi = \xi_0 \equiv t_{n-1}(x_{n-1} - x_{n-2}) + \cdots + t_1(x_1 - x_0) + x_0 \\
t_n &= t_{n-1} \Rightarrow \xi = \xi_1 \\
&\equiv t_{n-1}(x_n - x_{n-2}) + t_{n-2}(x_{n-2} - x_{n-3}) + \cdots + t_1(x_1 - x_0) + x_0
\end{aligned}$$

In this case, the intermediate integral in relation is as follows:

$$\begin{aligned}
&\int_0^{t_{n-1}} dt_n \times f^{(n)}(t_n(x_n - x_{n-1}) + \cdots + t_1(x_1 - x_0) + x_0) \\
&= \int_{\xi_0}^{\xi_1} f^{(n)}(\xi) \frac{d\xi}{x_n - x_{n-1}} = \frac{f^{(n-1)}(\xi_1) - f^{(n-1)}(\xi_0)}{x_n - x_{n-1}}
\end{aligned}$$

Using the inductive hypothesis in the relation, we obtain:

$$\begin{aligned}
&\int_0^1 dt_1 \int_0^{t_1} dt_2 \cdots \int_0^{t_{n-2}} dt_{n-1} \times \frac{f^{(n-1)}(\xi_1) - f^{(n-1)}(\xi_0)}{x_n - x_{n-1}} \\
&= \frac{f[x_0, \ldots, x_{n-2}, x_n] - f[x_0, \ldots, x_{n-2}, x_{n-1}]}{x_n - x_{n-1}} \\
&= f[x_0, \ldots, x_n]
\end{aligned}$$

1.3.2.6 Note

(1) If $f^{(n)}(x)$ is continuous on the interval $[a, b]$ and $y_0, \ldots, y_n$ are points in the $[a, b]$ and $x \in [a, b]$ is distinct from y_i, then

$$f[x, y_0, ..., y_n] = \frac{f[x, y_1, ..., y_n] - f[y_0, ..., y_n]}{x - y_0}$$

is a unique continuous generalization of the definition of divided differences.

(2) If $\{x_i\}$ and $\{y_i\}$ are two sets of variables in $[a, b]$ where $x_i \neq y_j, 0 \leq i \leq p, 0 \leq j \leq q, 0 \leq p, q \leq m$ and $f^{(m)}(x)$ is continuous on the interval $[a, b]$, then:

$$f\left[x_0, ..., x_p, y_0, ..., y_q\right] = g\left[x_0, ..., x_p\right] = h\left[y_0, ..., y_q\right]$$

where

$$g(x) \equiv f\left[x, y_0, ..., y_q\right], h(y) \equiv f\left[x_0, ..., x_p, y\right]$$

gives the unique continuous generalization of the definition of divided differences.

(3) If $f^{(n)}(x)$ is continuous in $[a, b]$ and $x_0, ..., x_n$ are points in $[a, b]$, then:

$$f[x_0, ..., x_n] = \frac{f^{(n)}(\xi)}{n!}, \xi \in I[x_0, ..., x_n]$$

1.3.2.7 Problem

If function $f(x)$ has continuous derivatives up to order m on the interval $[a, b]$ and $\{x_i\}$, $\{y_j\}$ and $\{z_k\}$ are sets of variables in $[a, b]$, so that $x_i \neq y_j, x_i \neq z_k, y_j \neq z_k$, $0 \leq i \leq p, 0 \leq j \leq q, 0 \leq k \leq r$ and $0 \leq p, q, r \leq m$, then prove

$$f\left[x_0, ..., x_p, y_0, ..., y_q, z_0, ..., z_r\right] = \frac{1}{p!q!r!}\left[\frac{\partial^p}{\partial y^q} \cdot \frac{\partial^q}{\partial y^q} \cdot \frac{\partial^r}{\partial z^r} f[x, y, z]\right]_{(\xi,\eta,\gamma)}$$

where

$$\gamma \in I[z_0, ..., z_r], \eta \in I\left[y_0, ..., y_q\right], \xi \in I\left[x_0, ..., x_p\right]$$

Solution: Suppose that:

$$\begin{aligned} g(x) &\equiv f\left[x, y_0, ..., y_q, z_0, ..., z_r\right] \\ h(y) &\equiv f[x, y, z_0, ..., z_r] \\ k(z) &\equiv f[x, y, z] \end{aligned}$$

Using the points (2) and (3) and their appropriate generalization for the set of variables $\{x_i\}$, $\{y_j\}$ and $\{z_k\}$, we will have:

$$g[x_0, \ldots, x_p] = \frac{1}{p!}\left[\frac{\partial^p}{\partial x^p} g(x)\right]$$

$$g(x) = h[y_0, \ldots, y_q] = \frac{1}{q!}\left[\frac{\partial^q}{\partial y^q} h(y)\right]_{y=\eta}$$

$$h(x) = k[z_0, \ldots, z_r] = \frac{1}{r!}\left[\frac{\partial^r}{\partial y^r} k(z)\right]_{z=\gamma}$$

According to the above relations, the proof is completed.

1.3.2.8 Problem

Suppose that $x_0, \ldots, x_n$ are mutual distinct points. Prove that the coefficients $a_0, \ldots, a_n$ in interpolation polynomials $p \in P_n$ are continuously dependent on the values of $y_0, \ldots, y_n$.

Solution: Suppose that

$$p(x) = a_0 + (x - x_0)a_1 + \cdots + (x - x_0)\cdots(x - x_{n-1})a_n$$

is the polynomial interpolation of the function f at the points $x_0, \ldots, x_n$. Therefore, given that we must have:

$$f(x_i) = p(x_i), i = 0, \ldots, n$$

a_i's can be calculated as:

$$\begin{aligned} f(x_0) &= p(x_0) = a_0 \\ f(x_1) &= p(x_1) = a_0 + a_1(x_1 - x_0) \\ &= f(x_0) + a_1(x_1 - x_0) \end{aligned}$$

So:

$$a_1 = \frac{f(x_1) - f(x_0)}{x_1 - x_0}$$

$$\begin{aligned} f(x_2) &= p(x_2) = a_0 + a_1(x_2 - x_0) + a_2(x_2 - x_0)(x_2 - x_1) \\ &= f(x_0) + \frac{f(x_1) - f(x_0)}{x_1 - x_0}(x_2 - x_0) + a_2(x_2 - x_0)(x_2 - x_1) \end{aligned}$$

Then:

$$a_2 = \frac{f(x_2) - f(x_0)}{(x_2 - x_0)(x_2 - x_1)} - \frac{f(x_1) - f(x_0)}{(x_1 - x_0)(x_2 - x_1)}$$

Given that $y_i \equiv f(x_i)$, therefore

$$\begin{aligned} a_0 &= y_0 \\ a_1 &= \frac{y_1 - y_0}{x_1 - x_0} \\ a_2 &= \frac{y_2 - y_0}{(x_2 - x_0)(x_2 - x_1)} - \frac{y_1 - y_0}{(x_1 - x_0)(x_2 - x_1)} \end{aligned}$$

Similarly, by continuing the same process, the other a_i is obtained in terms of the points and y_i.

1.3.2.9 Note

The operator Δ is a forward operator and is defined as follows:

$$\Delta f_i = f_{i+1} - f_i$$

1.3.2.10 Problem

Prove that

$$\Delta(f(x) \cdot g(x)) = f(x) \cdot \Delta g(x) + \Delta f(x) \cdot g(x+h)$$

Solution:

$$\begin{aligned} \Delta(f(x) \cdot g(x)) &= f(x+h) \cdot g(x+h) - f(x) \cdot g(x) \\ &= f(x+h) \cdot g(x+h) - f(x) \cdot g(x) \\ &\quad + f(x) \cdot g(x+h) = f(x) \cdot g(x+h) \\ &= g(x+h)(f(x+h) - f(x)) + f(x)(g(x+h) - g(x)) \\ &= g(x+h) \cdot \Delta f(x) + f(x) \cdot \Delta g(x) \end{aligned}$$

1.3.2.11 Problem

If $f(x) = \frac{1}{x+c}$, where c is a constant (real or complex) number, prove that:

$$f[x_0, \ldots, x_n] = \frac{(-1)^n}{(x_0 + c) \cdots (x_n + c)}$$

Solution: We prove it by induction on n.

Initial case: Suppose $n = 1$:

$$\begin{aligned} f[x_0, x_1] &= \frac{f(x_1) - f(x_0)}{x_1 - x_0} \\ &= \frac{\frac{1}{x_1+c} - \frac{1}{x_0+c}}{x_1 - x_0} \\ &= \frac{x_0 - x_1}{(x_0 + c)(x_1 + c)} \cdot \frac{1}{x_1 - x_0} \\ &= \frac{-1}{(x_0 + c)(x_1 + c)} \end{aligned}$$

Induction hypothesis: Suppose that the above relation is hold for $n = k$.

Induction step: We prove that it holds for 2 as well.

According to the induction hypothesis, we know:

$$\begin{aligned} f\big[x_1, ..., x_{k+1}\big] &= \frac{(-1)^k}{(x_1 + c)...(x_{k+1} + c)} \\ f[x_0, ..., x_k] &= \frac{(-1)^k}{(x_0 + c)...(x_k + c)} \end{aligned}$$

So

$$\begin{aligned} f\big[x_0, ..., x_k, x_{k+1}\big] &= \frac{f\big[x_1, ..., x_{k+1}\big] - f[x_0, ..., x_k]}{x_{k+1}, x_0} \\ &= \left(\frac{(-1)^k}{(x_1 + c)...(x_{k+1} + c)} = \frac{(-1)^k}{(x_0 + c)...(x_k + c)}\right) \\ &\quad \times \left(\frac{1}{x_{k+1} - x_0}\right) \\ &= \frac{(-1)^k(x_0 - x_{k+1})}{(x_0 + c)...(x_{k+1} + c)} \cdot \frac{1}{x_{k+1} - x_0} \\ &= \frac{(-1)^{k+1}}{(x_0 + c)...(x_{k+1} + c)} \end{aligned}$$

Therefore, the proof is completed.

1.3.2.12 Problem

Suppose that $x_0, ..., x_n$ are distinct points, and $f \in c^n(I(x_0, ..., x_n))$, in this case:

$$f[x_0, ..., x_n] = \int_{\tau n} ... \int f^{(n)}(t_0x_0 + \cdots + t_nx_n)dt_1dt_2...dt_n$$

where are $t_0 = 1 - \sum_{j=1}^{n} t_j$ and $\tau_n = \left\{(t_0, ..., t_n)|t_j \geq 0, \sum_{j=1}^{n} t_j \leq 1\right\}$.

Solution: We prove the assertion by induction on n.

Initial case: Suppose $n = 1$ with $\tau_1 = [0, 1]$ and $t_0 = 1 - t_1$, then

$$\begin{aligned}\int_{\tau_1} f'(t_0x_0 + t_1x_1)dt_1 &= \int_0^1 f'(x_0 + t_1(x_1 - x_0))dt_1 \\ &= \frac{f(x_1) - f(x_0)}{x_1 - x_0} \\ &= f[x_0, x_1]\end{aligned}$$

Induction hypothesis: Assume that the assertion holds for $n = k - 1$.

Induction step: We prove that the sentence holds for $n = k$ as well. Note that $t_0 = 1 - t_1 - \cdots - t_k$ and $\tilde{t}_k = 1 - t_1 - \cdots - t_{k-1}$ so:

$$\begin{aligned}&\int_{\tau_k} ... \int f^{(k)}(t_0x_0 + \cdots + t_kx_k)dt_1 \cdots dt_k \\ &= \int_{\tau_{k-1}} ... \int \int_0^{\tilde{t}_k} f^{(k)}(x_0 + t_1(x_1 - x_0) + \cdots + t_k(x_k - x_0))dt_kdt_1 \cdots dt_{k-1} \\ &\int_{\tau_{k-1}} ... \int \frac{1}{x_k - x_0}\Big[f^{(k-1)}\big(x_0 + t_1(x_1 - x_0) + \cdots + \tilde{t}_k(x_k - x_0)\big) \\ &\quad - f^{(k-1)}\big(\tilde{t}_kx_0 + t_1x_1 + \cdots + t_{k-1}x_{k-1}\big)\Big]dt_1 \cdots dt_{k-1} \\ &= \int_0^1 \int_0^{1-t_1} ... \int_0^{1-t_1-\cdots-t_{k-2}} \frac{f^{(k-1)}\big(t_1x_1 + \cdots + \tilde{t}_kx_k\big)}{x_k - x_0}dt_1 \cdots dt_{k-1} \\ &\quad - \int ... \int \frac{f^{(k-1)}\big(\tilde{t}_kx_0 + t_1x_1 + \cdots + t_{k-1}x_{k-1}\big)}{x_k - x_0}dt_1...dt_{k-1} \\ &= \frac{f[x_1, ..., x_k] - f\big[x_0, ..., x_{k-1}\big]}{x_k - x_0} \\ &= f[x_0, ..., x_k]\end{aligned}$$

1.3.2.13 Problem

Suppose $x_0 = x_1 = \cdots = x_n$. Show using the assumptions of problem (Sect. 1.3.2.12):

$$\lim_{x_j \to x_0} f[x_0, ..., x_n] = \frac{f^{(n)}(x_0)}{n!}$$

Solution:

$$\begin{aligned}
\lim_{x_j \to x_0} f[x_0, ..., x_n] &= \int ... \int_{\tau_n} f^{(n)}((t_0 + \cdots + t_n)x_0) dt_1 \cdots dt_n \\
&= \int ... \int_{\tau_n} f^{(n)}(x_0) dt_1 \cdots dt_n \\
&= f^{(n)}(x_0) \int ... \int_{\tau_n} dt_1 \cdots dt_n \\
&= \frac{f^{(n)}(x_0)}{n!}
\end{aligned}$$

1.3.2.14 Problem

Show that if α is constant and positive, and $f(x) = \alpha^x$, then in what conditions

$$\Delta f_i = f_i$$

always holds.

Solution: We have to find the condition for $\Delta f_i = f_{i+1} - f_i = f_i$.

$$\begin{aligned}
& f_{i+1} = 2 f_i \\
& \Rightarrow \alpha^{x_i + 1} = 2\alpha^{x_i} \\
& \Rightarrow \ln \alpha^{x_i + 1} = \ln 2\alpha^{x_i} \\
& \Rightarrow x_{i+1} \ln \alpha = \ln 2 + x_i \ln \alpha \\
& \Rightarrow (x_{i+1} - x_i) \ln \alpha = \ln 2 \\
& \Rightarrow x_{i+1} = x_i + \frac{\ln 2}{\ln \alpha}
\end{aligned}$$

So, by choosing $h = \frac{\ln 2}{\ln \alpha}$, $\Delta f_i = f_i$ will always hold.

1.4 A Short Review on Vector Norms and Linear System of Equations

Now we are going to review some of the basic concepts of vector norms and linear system of equation which are essential for a deeper understanding of intervals.

1.4.1 *Vector Norm*

In this section, first we introduce the norm of a real-valued vector and then point out its application in finding the bounds of errors and its approximate solution.

Usually, the norm of a vector is used to define a distance in $\mathbb{R}^n$.

1.4.1.1 Definition

A vector norm on $\mathbb{R}^n$ is a function from $\mathbb{R}^n$ to R that satisfies the following properties:

$$\|\cdot\| : \mathbb{R}^n \to \mathbb{R}$$

(1)

$$\forall X\big(X \in \mathbb{R}^n \Rightarrow \|X\| \geq 0\big)$$

(2)

$$\forall X\big(X \in \mathbb{R}^n \Rightarrow (X = (0, \ldots, 0) = 0 \iff \|X\| = 0)\big)$$

(3)

$$\forall X \forall \lambda\big((X \in \mathbb{R}^n \& \lambda \in \mathbb{R}\big) \Rightarrow \|\lambda X\| = |\lambda| \cdot \|X\|\big)$$

(4)

$$\forall X \forall Y\big((X \in \mathbb{R}^n \& Y \in \mathbb{R}\big) \Rightarrow \|X + Y\| \leq \|X\| + \|Y\|\big)$$

1.4.1.2 Definition

If a function does not hold property number (2) of those mentioned above, then, it is called semi-norm.

1.4.1.3 Definition

For a vector $X = (x_1, ..., x_n)^t$, the norms are defined as follows:

(1) Norm L_1

$$\|X\|_1 = \sum_{i=1}^{n} |x_i|$$

(2) Norm L_2

$$\|X\|_2 = \left(\sum_{i=1}^{n} |x_i|^2\right)^{\frac{1}{2}}$$

where norm L_2 is also called Euclidean norm.

(3) Norm L_p

$$\|X\|_p = \left(\sum_{i=1}^{n} |x_i|^p\right)^{\frac{1}{p}}$$

(4) Norm L_∞

$$\|X\|_\infty = \max_i |x_i|$$

Another important norm, also called elliptic norm, is as follows

$$\|X\| = \left(X^t B X\right)^{\frac{1}{2}} \tag{1.16}$$

where B is a definite positive symmetric matrix and satisfies the norm properties.

All defined norms are on space $\mathbb{R}^n$. In the case, H^t, H^H are Hermitian and B is the definite positive Hermitian $\left(B = \overline{B^t}\right)$ matrix, norm (1.16) is defined on space $\mathbb{C}^n$ and is easily proved to be an inner product. The following is the definition of inner product.

1.4.1.4 Definition

Function $\mathbb{R}^n \times \mathbb{R}^n \to \mathbb{R}$ defines inner product with the following properties:

(1) $\forall X\left(X \in \mathbb{R}^n \Rightarrow (X, X) \geq 0\right)$

(2) $\forall X\left(X \in \mathbb{R}^n \Rightarrow (X = 0 \iff (X, Y) = 0)\right)$

(3) $\forall X \forall Y \forall \alpha \big((X, Y \in \mathbb{R}^n \& \alpha \in \mathbb{R}) \Rightarrow (\alpha X, Y) = \alpha (X, Y) \big)$

(4) $\forall X \forall Y \big(X, Y \in \mathbb{R}^n \Rightarrow (X, Y) = (Y, X) \big)$

(5) $\forall X \forall Y \forall Z \big(X, Y, Z \in \mathbb{R}^n \Rightarrow (X + Y, Z) = (X, Z) + (Y, Z) \big)$

If we consider the intended space as a $\mathbb{C}^n \times \mathbb{C}^n$ complex space, the only difference that arises is that property (4) is defined as follows:

$$(X, Y) = \overline{(Y, X)}$$

Now if we assume $X = Y$, then for each inner product, a norm is defined as follows:

$$\|X\| = (X, X)^{\frac{1}{2}} \tag{1.17}$$

We should only prove the triangle inequality property. For this purpose, given the Cauchy–Schwarz inequality we have already proved, we have:

$$|(X, Y)|^2 \leq (X, X)(Y, Y)$$

Therefore

$$\begin{aligned} \|X + Y\|^2 &= (X, X) + 2(X, Y) + (Y, Y) \\ &\leq (X, X) + 2(X, X)^{\frac{1}{2}}(Y, Y)^{\frac{1}{2}} + (Y, Y) \\ &= (\|X\| + \|Y\|)^2 \end{aligned}$$

As a result

$$\|X + Y\| \leq \|X\| + \|Y\|$$

Now, considering Eqs. (1.16) and (1.17), for a definite positive symmetric matrix B, the inner product of two vectors X and Y is defined as follows:

$$(X, Y) = X^t BY$$

To prove that Eq. (1.16) is a norm, it is sufficient to prove that it satisfies the properties of inner product, and since the inner product defines a norm, Eq. (1.16) defines a vector norm.

If we want to have a two-dimensional image of the defined norms, we need to consider a unit sphere to which the norms are constrained. Level $\{X \mid \|X\| = 1\}$ is

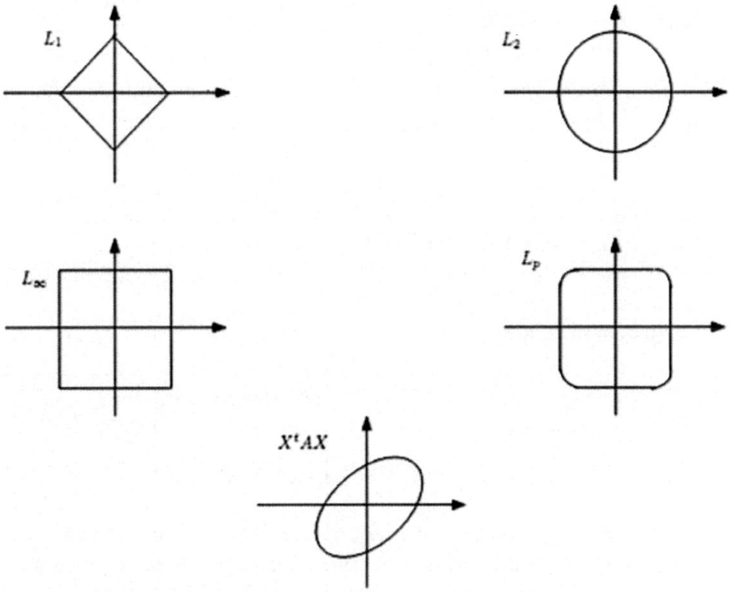

Fig. 1.1 Geometrical representation on norms

called unit sphere and set $\{X | \|X\| \leq 1\}$ is called unit ball where each norm on this ball is represented by the following shapes (Fig. 1.1).

These mentioned norms are theoretically useful to produce other type of norms. After defining the norm and introducing their types, we will discuss it further by providing some theorems.

The following theorem is known as the norm equivalence theorem. Given this theorem, it can be claimed that if a vector converges under one norm, it also converges under other norms. That is, if a sequence of vectors converges under norm L_∞, it converges under other norms such as L_1, L_2 and L_p.

1.4.1.5 Theorem (Equivalence of Norms)

Suppose $\|\cdot\|'$ *and* $\|\cdot\|$ *are two arbitrary norms on* $\mathbb{R}^n$ *or* $\mathbb{C}^n$*. In this case, there are constants* $0 < c_1 \leq c_2$ *such that for each vector X,*

$$c_1\|X\| \leq \|X\|' \leq c_2\|X\|$$

1.4.2 *Direct Methods*

Several numerical methods for solving the integral equations use the system of linear equations. To solve these systems, there are two types of methods, direct and iterative methods.

1.4.2.1 Gaussian Elimination Method

This method is one of the direct methods.

Consider the augmented matrix that for every $b_i := a_{i,n+1}$ $i = 1, ..., n$ as follows:

$$[A : b] = \begin{bmatrix} a_{11} & \cdots & a_{1,n+1} \\ \vdots & \ddots & \vdots \\ a_{n1} & \cdots & a_{n,n+1} \end{bmatrix}$$

In this method, we use a procedure that converts the matrix above into an upper triangular matrix with nonzero diagonal elements. These conversions are performed according to the matrix rank discussed in the previous section.

The first step in the Gaussian elimination method is to set elements below a_{11} equal to zero, provided that $a_{11} \neq 0$. If $a_{11} = 0$, given that $det(A) \neq 0$, then in the first column, there is a row, for example pth row, where $a_{p1} \neq 0$, because otherwise, $det(A) = 0$, which is a contradiction. Now we interchange the first and pth rows so that the $(1, 1)$ element of the matrix is nonzero. Then, we add $\left(\frac{-a_{i1}}{a_{11}}\right)$-times of the first row to the second to nth rows, denoting the new coefficients by $a_{ij}^{(1)}$.

If we assume $m_{i1} = \frac{-a_{i1}}{a_{11}}$ for $i = 1, ..., n$, we have:

$$a_{ij}^{(1)} = a_{ij} - \frac{a_{i1}}{a_{11}} a_{1j} = a_{ij} + m_{i1} \cdot a_{1j}$$
$$i = 2, ..., n,\ j = 2, ..., n+1$$

The resulting matrix is as follows

$$\begin{bmatrix} a_{11} & a_{12} & \cdots & a_{1n} & : a_1, n+1 \\ 0 & a_{22}^{(1)} & \cdots & a_{2n}^{(1)} & : a_{2,n+1}^{(1)} \\ \vdots & \vdots & \vdots & \vdots & \vdots \\ 0 & a_{n2}^{(1)} & \cdots & a_{nn}^{(1)} & : a_{n,n+1}^{(1)} \end{bmatrix} \tag{1.18}$$

The second step of the Gaussian elimination method is to set the elements below $a_{22}^{(1)}$ equal to zero, which according to the discussion of the first step, $a_{22}^{(1)} \neq 0$. Otherwise, there must be an index of rows like r, where $3 \leq r \leq n$ and $a_{r2}^{(1)} \neq 0$. Because if it is not the case, the first and second columns are multiplications of each other, so $det(A) = 0$ that is a contradiction. Then $a_{r2}^{(1)} \neq 0$, where the second and rth rows can be interchanged, and the (2, 2) element of the matrix (1.18) becomes nonzero. We add $\left(\frac{-a_{i2}^{(1)}}{a_{22}^{(1)}}\right)$ times of the second row to the third to nth rows. The new coefficients are shown as $a_{ij}^{(2)}$. If we assume that $m_{i2} = \frac{-a_{i2}^{(1)}}{a_{22}^{(1)}}, i = 3, ..., n$ for $i = 3, ..., n$, we have:

$$a_{ij}^{(2)} = a_{ij}^{(1)} - \frac{a_{i2}^{(1)}}{a_{22}^{(1)}} \cdot a_{2j}^{(1)} = a_{ij}^{(1)} + m_{i2} \cdot a_{2j}^{(1)}$$
$$i = 3, ..., n,\ j = 3, ..., n+1$$

The resulting matrix is as follows

$$\begin{bmatrix} a_{11} & a_{12} & a_{13} & \cdots & a_{1n} & : a_{1,n+1} \\ 0 & a_{22}^{(1)} & a_{23}^{(1)} & \cdots & a_{2n}^{(1)} & : a_{2,n+1}^{(1)} \\ 0 & 0 & a_{33}^{(2)} & \cdots & a_{3n}^{(2)} & : a_{3,n+1}^{(2)} \\ \vdots & \vdots & \vdots & & \vdots & \vdots \\ 0 & 0 & a_{n3}^{(2)} & \cdots & a_{nn}^{(2)} & : a_{n,n+1}^{(2)} \end{bmatrix} \tag{1.19}$$

We continue the same process until reaching the following matrix after $n-1$ step:

$$\begin{bmatrix} a_{11} & a_{12} & a_{13} & \cdots & a_{1n} & : a_{1,n+1} \\ 0 & a_{22}^{(1)} & a_{23}^{(1)} & \cdots & a_{2n}^{(1)} & : a_{2,n+1}^{(1)} \\ 0 & 0 & a_{33}^{(2)} & \cdots & a_{3n}^{(2)} & : a_{3,n+1}^{(2)} \\ \vdots & \vdots & \vdots & \ddots & \vdots & \vdots \\ 0 & 0 & 0 & \cdots & a_{nn}^{(n-1)} & : a_{n,n+1}^{(n-1)} \end{bmatrix} \tag{1.20}$$

All of the diagonal coefficients are nonzero. In general, the coefficients are obtained as follows:

$$a_{ij}^{(k)} = a_{ij}^{(k-1)} + m_{ik} \cdot a_{kj}^{(k-1)},\ k = 1, ..., n-1$$
$$m_{ik} = -\frac{a_{ik}^{(k-1)}}{a_{kk}^{(k-1)}},\ i = k+1, ..., n$$
$$a_{ij}^{(0)} = a_{ij},\ j = k+1, ..., n+1 \tag{1.21}$$

The system of equations that is obtained from the last step is as follows:

$$a_{11}x_1 + a_{12}x_2 + a_{13}x_3 + \cdots + a_{1n}x_n = a_{1,n+1}$$
$$a_{22}^{(1)}x_2 + a_{23}^{(1)}x_3 + \cdots + a_{2n}^{(1)}x_n = a_{2,n+1}^{(1)}$$
$$a_{33}^{(2)}x_3 + \cdots + a_{3n}^{(2)}x_n = a_{3,n+1}^{(2)}$$

$$\ddots$$

$$a_{nn}^{(n-1)} x_n = a_{n,n+1}^{(n-1)} \tag{1.22}$$

This system is solved by backward substitution, so that

$$x_i = \frac{1}{a_{ii}^{(i-1)}} \left[a_{in+1}^{(i-1)} - \sum_{j=i+1}^{n} a_{ij}^{(i-1)} x_j \right], i = 1, \dots, n \tag{1.23}$$

1.4.2.2 Remark

If the Gaussian elimination method can be performed on the system $AX = b$ without interchanging the rows, then the matrix A is decomposed by multiplying a lower triangular matrix into an upper triangular matrix:

$$A = LU$$

where $U = (u_{ij})$ and $L = (l_{ij})$ are defined as follows:

$$u_{ij} = \begin{cases} a_{ij}^{(i)}, i = 1, \dots, j-1 \\ 0, i = j+1, \dots, n \end{cases}$$

$$l_{ij} = \begin{cases} 0, i = 1, \dots, j-1 \\ 1, i = j \\ m_{ij}, i = j+1, \dots, n \end{cases}$$

Therefore,

$$det(A) = det(LU) = det(L) \cdot det(U) = \prod_{i=1}^{n} u_{ii} \times \prod_{i=1}^{n} l_{ii}$$

So $det(A) \neq 0$.

The non-singularity of the matrix A is a necessary condition for the LU decomposition, but it is not a sufficient condition.

1.4.2.3 *LU*Decomposition of the Matrix

In this section, it is tried to convert matrix A to a product of the lower triangular matrix L and the upper triangular matrix U. This decomposition has many applications in solving the system of linear equations. The decomposition is performed in two ways, one using the Gaussian elimination method and the other using direct decomposition, which is discussed in the following.

1.4.2.4 The *LU* Decomposition

If $det A \neq 0$, there is a permutation matrix such as P so that PA can be decomposed into LU.

Consider the lower triangular matrix L_i as follows.

$$L_i = \begin{bmatrix} 1 & & & & & \\ & \ddots & & & 0 & \\ & & 1 & & & \\ & & m_{i+1,i} & 1 & & \\ & 0 & \vdots & & \ddots & \\ & & m_{n,i} & & & 1 \end{bmatrix} \tag{1.24}$$

If the matrix A is as follows:

$$\begin{bmatrix} a_{11} & a_{12} & \cdots & a_{1i} & \cdots & a_{1n} \\ \vdots & \vdots & & \vdots & & \vdots \\ a_{i1} & a_{i2} & \cdots & a_{ii} & \cdots & a_{in} \\ a_{i+1,1} & a_{i+1,2} & \cdots & a_{ii} & \cdots & a_{i+1,n} \\ \vdots & \vdots & & \vdots & & \vdots \\ a_{n1} & a_{n2} & \cdots & a_{ni} & \cdots & a_{nn} \end{bmatrix}$$

In this case,

$$L_i A = \begin{bmatrix} a_{11} & a_{12} & \cdots & a_{1i} & \cdots & a_{1n} \\ \vdots & \vdots & & \vdots & & \vdots \\ a_{i1} & a_{i2} & \cdots & a_{ii} & \cdots & a_{in} \\ a_{i+1,1}^{(1)} & a_{i+1,2}^{(1)} & \cdots & 0 & \cdots & a_{i+1,n}^{(1)} \\ \vdots & \vdots & & \vdots & & \vdots \\ a_{n1} & a_{n2} & \cdots & 0 & \cdots & a_{nn}^{(1)} \end{bmatrix}$$

That is, by multiplying into L_i, the elements below a_{ii} become zero. Suppose that $L_i^{-1} = \widehat{L}_i$. The matrix $\widehat{L}_i$ is the same as the matrix L_i, with only one exception that the signs of m_{ij}'s are changed.

The obtained matrix from the application of the first step of the Gaussian elimination method on the matrix of coefficients A can be written as follows:

$$A_1 = L_1 P_{1r_1} A, r_1 \geq 1$$

That is, the first and second rows are interchanged before removing. (If no interchange is required, i.e., $a_{11} \neq 0$, then $r_1 = 1$ and $P_{1,r_1} = I$).

In general, if we show the coefficient matrix in the ith step of the Gaussian elimination method by A_i, then

$$A_i = L_i P_{i,r_i} A_{i-1}, r_i \geq i, i = 1, ..., n-1$$

Before the i'st elimination stage, rth and r_ith rows are interchanged. Continuing the same process in the $(n-1)$th step, the upper triangular coefficient matrix A_{n-1} in the system of equations (3.17) will be as $U = [u_{ij}]$.

$$U = A_{n-1} = L_{n-1} P_{n-1,r_{n-1}} L_{n-2} P_{n-2,r_{n-2}} \cdots L_2 P_{2,r_2} L_1 P_{1,r_1} A \tag{1.25}$$

If matrix P_{i,r_i} was already known, we would have to use it for matrix A from the beginning.

$$\tilde{A} = P_{n-1,r_{n-1}} P_{n-2,r_{n-2}} ... P_{1,r_1} A = PA$$

where P is the permutation matrix in the form of $P = P_{n-1,r_{n-1}} ... P_{1,r_1}$ and we will be able to perform the Gaussian elimination method without interchanging the rows, as a result, Eq. (1.25) will be written as follows

$$U = L_{n-1} \cdots L_2 L_1 (P_{n-1,r_{n-1}} \cdots P_{1,r_1}) A = \tilde{L} \tilde{A} \tag{1.26}$$

where

$$\tilde{L} := L_{n-1} L_{n-2} \cdots L_2 L_1$$

which is a product of a finite number of lower triangular matrices; a lower triangular matrix is as follows:

$$\tilde{L} = \begin{bmatrix} 1 & & & & \\ m_{21} & 1 & & 0 & \\ m_{31} & m_{32} & \ddots & & \\ \vdots & & & 1 & \\ m_{n1} & m_{n2} & & m_{n,n-1} & 1 \end{bmatrix}$$

as a result

$$\tilde{L}^{-1} = L_1^{-1} L_2^{-1} \cdots L_{n-1}^{-1} := L$$

which is as follows

$$\tilde{L} = \begin{bmatrix} 1 & & & & \\ -m_{21} & 1 & & 0 & \\ -m_{31} & -m_{32} & \ddots & & \\ \vdots & & & 1 & \\ -m_{n1} & -m_{n2} & & -m_{n,n-1} & 1 \end{bmatrix}$$

So we have

$$\tilde{A} = PA = LU$$

So we showed that if A is a non-singular matrix, then there is a permutation matrix such as P, so that $PA = \tilde{A}$ is a product of the upper triangular matrix U and the lower triangular matrix L. This product is called the LU decomposition of the matrix $\tilde{A}$.

This decomposition is unique. Because if we assume $\tilde{A} = L_1U_1$ where L_1 is a lower triangular matrix and U_1 is an upper triangular matrix.

$$LU = L_1U_1$$

In this case $UU_1^{-1} = L^{-1}L_1$ because U_1 and L are non-singular. Since the inverse of an identity lower triangular matrix (on the diagonal, one) has the same form and the product of the two lowers triangular (identity) matrices has also the same form, an upper triangular matrix is equal to a lower triangular matrix, and this is possible if both matrices are identity matrices. i.e.,

$$U_1 = U, L_1 = L$$

Now we prove by induction that the decomposition of the matrix A as LU is singular. Suppose $u_{ii} = 1, i = 1, ..., n$.

$a_{ij} = \sum_{r=1}^{\min(i,j)} l_{ir}u_{rj}$: ith row and jth column.

$$A = \begin{bmatrix} l_{11} & & & & & \\ l_{21} & l_{22} & & & 0 & \\ \vdots & \vdots & \ddots & & & \\ l_{k1} & l_{k2} & \cdots & l_{kk} & & \\ \vdots & \vdots & & \vdots & \ddots & \\ l_{n1} & l_{n2} & \cdots & l_{nk} & \cdots & l_{nn} \end{bmatrix} \begin{bmatrix} 1 & u_{12} & \cdots & u_{1k} & \cdots & u_{1n} \\ 0 & 1 & \cdots & u_{2k} & \cdots & u_{2n} \\ & & \ddots & \vdots & & \vdots \\ & 0 & & 1 & & u_{kn} \\ & & & & \ddots & \vdots \\ & & & & & 1 \end{bmatrix}$$

In each step, we uniquely specify a column of L and a row of U.

$$a_{i1} = l_{i1}, i = 1, \ldots, n \tag{1.27}$$

The first column of L is the same as the first column of A.

$$a_{1j} = l_{11}u_{1j} \Rightarrow u_{1j} = \frac{a_{1j}}{l_{11}} = \frac{a_{1j}}{a_{11}}, l_{11} \neq 0, j = 2, \ldots, n \tag{1.28}$$

Therefore, the first row of U is uniquely specified.

Induction hypothesis: Suppose the first to $(k-1)$th columns of L and the first to $(k-1)$th rows of U are uniquely specified.

Induction step: kth column of L and kth row of U are uniquely specified.

We have:

$$a_{ij} = \sum_{r=1}^{\min(i,j)} l_{ir}u_{rj}$$

The goal is to specify $l_{kk}, l_{k+1,k}, \ldots, l_{nk}$.

Suppose $i \geq k$, so $\min(i, k) = k$. Then:

$$a_{ik} = \sum_{r=1}^{k} l_{ir}u_{rk} = \sum_{r=1}^{k-1} l_{ir}u_{rk} + l_{ik}$$

$$l_{ik} = a_{ik} = \sum_{r=1}^{k-1} l_{ir}u_{rk}, i = k, k+1, \ldots, n \tag{1.29}$$

According to the inductive hypothesis, $\sum_{r=1}^{k-1} l_{ir}u_{rk}$ is uniquely specified. So $i = k, k+1, \ldots, n$, l_{ik}, that are the elements of the kth column of L are uniquely identified. Now, we uniquely specify $u_{k,k+1}, \ldots, u_{kn}$.

Suppose $k \leq j$, so $\min(i, j) = k$. We have:

$$a_{kj} = \sum_{r=1}^{k} l_{kr}u_{rj} = \sum_{r=1}^{k-1} l_{kr}u_{rj} + l_{kk}u_{kj}, j = k, k+1, \ldots, n$$

$\sum_{r=1}^{k-1} l_{kr}u_{rj}$ is uniquely specified according to the induction hypothesis, and since all the forward submatrices are non-singular, then $l_{kk} \neq 0$ and it is known according to the previous case.

$$u_{kj} = \frac{1}{l_{kk}}\left(a_{kj} - \sum_{r=1}^{k-1} l_{kr}u_{rj}\right), j = k+1, \ldots, n \tag{1.30}$$

Therefore, the kth row of U is also uniquely specified.

Because of the importance of computational complexity in numerical algorithms, it is essential to have an estimate on the complexity, so in this algorithm, $(n-1)$ multiplication operations can be counted in Eq. (1.28) and

$$\sum_{k=2}^{n}(n-k+)(k-1) \simeq \frac{n^3}{6}$$

multiplication operations can be counted in Eq. (1.29) and

$$\sum_{k=2}^{n}(n-k)k \simeq \frac{n^3}{6}$$

multiplication operations can be counted in Eq. (1.30) and therefore a total of

$$\frac{n^3}{6} + \frac{n^3}{6} + (n-1) \simeq \frac{n^3}{3}$$

multiplication operations can be counted in the algorithm, so it can be said that the complexity is of the order of about $O\left(\frac{n^3}{3}\right)$.

The application of the decomposition $A = LU$ in the form of $u_{ii} = 1, i = 1, ..., n$ to solve the system $AX = b$ is as follows:

$$AX = b \Rightarrow (LU)X = b \Rightarrow L(UX) = b$$

If we assume $UX = y$, then $Ly = b$, where first the first system is solved by backward substituting and the second system is solved by forward substituting. For solving the system $Ly = b$, we have:

$$y_1 = \frac{b_1}{l_{11}}, l_{11} \neq 0$$
$$y_i = \frac{1}{l_{ii}}\left(b_i - \sum_{r=1}^{i-1} l_{ir} y_r\right), i = 2, ..., n$$

The number of the multiplication and division operations is:

$$1 + \sum_{i=2}^{n} i = \frac{n(n+1)}{2}$$

Similarly, the number of multiplication and division operations in solving the system $UX = y$ is $\frac{n(n+1)}{2}$. So, the total number of three operations is of order $O\left(n^2\right)$.

1.4.2.5 Cholesky Decomposition

Now, suppose $E^2 = D$ in the equation $A = LDL^t$, because:

$$D = \begin{bmatrix} d_1 & & 0 \\ & \ddots & \\ 0 & & d_n \end{bmatrix}$$

So,

$$E = \begin{bmatrix} \sqrt{d_1} & & 0 \\ & \ddots & \\ 0 & & \sqrt{d_n} \end{bmatrix}$$

Therefore, for the matrix A we have:

$$A = LE^2L^t = LE \cdot EL^t = (LE)(LE)^t$$

And if we assume that $LE = M$, then M is a lower triangular matrix, then

$$A = MM^t$$

In this decomposition, $d_i > 0$ $(i = 1, ..., n)$ is necessary. This decomposition is called Cholesky decomposition. Now we express the Cholesky decomposition of the symmetric matrix A structurally by using induction. As before, we have:

$$A_k = M_k M_k^t \Rightarrow A_{k+1} = M_{k+1} M_{k+1}^t$$

where

$$A_{k+1} = \begin{bmatrix} A_k & C_{k+1} \\ C_{k+1}^t & a_{k+1,k+1} \end{bmatrix}, M_{k+1} = \begin{bmatrix} M_k & 0 \\ m_{k+1}^t & m_{k+1,k+1} \end{bmatrix}$$

The unknowns are:

$$m_{k+1,k+1}, m_{k+1}^t$$

By substitution, we have:

$$M_k \cdot m_{k+1} = C_{k+1}$$

So, m_{k+1} is calculated and finally

$$(m_{k+1,k+1})^2 = a_{k+1,k+1} - m_{k+1}^t m_{k+1}$$

1.4.2.6 Example

To solve the system

$$\begin{bmatrix} 4 & 2 & -2 \\ 2 & 2 & -3 \\ -2 & -3 & 14 \end{bmatrix} \begin{bmatrix} x_1 \\ x_2 \\ x_3 \end{bmatrix} = \begin{bmatrix} 10 \\ 5 \\ 4 \end{bmatrix}$$

Given that the matrix of coefficients is symmetric, we have

$$M = LE = \begin{bmatrix} 2 & 0 & 0 \\ 1 & 1 & 0 \\ -1 & -2 & 3 \end{bmatrix}$$

Why?

$$AX = b \Rightarrow (MM^t)X = b \Rightarrow M(M^tX) = b$$

If we assume that $M^tX = C$, then $MC = b$. Therefore,

$$C = \begin{bmatrix} 5 \\ 0 \\ 3 \end{bmatrix}$$

And as a result

$$M^tX = C \Rightarrow X = \begin{bmatrix} 2 \\ 2 \\ 1 \end{bmatrix}$$

1.4.2.7 Theorem

The matrix A has a Cholesky decomposition if and only if A is a positive definite matrix.

Proof Suppose that A has a Cholesky decomposition, then $A = LL^t$ where L is a non-singular lower triangular matrix.

$$\forall X(X \neq 0 \Rightarrow X^tAX = X^tLL^tX = Y^tY)$$

where $L^t X = Y$. Because L^t is non-singular, so for every $X \neq 0$, Y is nonzero, therefore:

$$X^t AX = Y^t Y = \sum y_i^2 > 0$$

That is, A is positive definite.

Now suppose that A is positive definite, therefore:

$$A = A^t \& \forall X = \left(X \neq 0 \Rightarrow X^t AX > 0\right)$$

it can be written:

$$0 < X^t AX = X^t LDL^t X = Y^t DY = \sum y_i^2 d_{ii}$$

where $L^t X = Y$. Because L^t is non-singular, so there is a nonzero X so that $L^t X = Y = e_i$, therefore $d_{ii} > 0$.

Since the elements of D are positive, then we can write:

$$A = LD^{\frac{1}{2}} D^{\frac{1}{2}} L^t = MM^t$$

where $M = LD^{\frac{1}{2}}$ is a lower triangular matrix. So, the matrix A has the Cholesky decomposition.

In this case, $D \geq 0$.

Because A is a M matrix, $a_{ii} > 0$ and therefore,

$$A + D = A - \lambda I$$

is a M matrix, so $(A - \lambda I)^{-1}$ exists, therefore:

$$det(A - \lambda I) \neq 0 \tag{1.31}$$

Equation (1.31) is contradictory, so the hypothesis is false and A is a positive definite matrix.

We now state a method called successive over relaxation (SOR) to obtain the convergence for the systems that do not converge by the Gauss–Seidel method.

1.4.3 Numerical Methods

In this section, we discuss three types of simple iterative methods. These methods are not generally convergent but in convergence cases, the SOR methods have some advantages over the others.

1.4.3.1 Jacobi's Iterative Method

Suppose that $det(A) \neq 0$, then A can be converted to a matrix with nonzero diagonal elements using permutation matrices.

$$\begin{aligned} A &= D - (L + U) \\ &= D - L - U \end{aligned}$$

where

$$D = \begin{bmatrix} a_{11} & 0 & \cdots & 0 \\ 0 & a_{22} & 0 & 0 \\ \vdots & \vdots & \ddots & \vdots \\ 0 & 0 & \cdots & a_{nn} \end{bmatrix}$$

$$L = \begin{bmatrix} 0 & & & 0 \\ -a_{21} & 0 & & \\ \vdots & \vdots & \ddots & \\ -a_{n1} & -a_{n2} & \cdots & -a_{n,n-1} & 0 \end{bmatrix}$$

$$U = \begin{bmatrix} 0 & -a_{12} & \cdots & -a_{1n} \\ & 0 & & \vdots \\ 0 & & \ddots & -a_{n-1,n} \\ & & & 0 \end{bmatrix}$$

We have

$$[D - (L + U)]X = b$$

Then,

$$DX = (L + U)X + b$$

as a result

$$X = D^{-1}(L + U)X + D^{-1}b$$

Then,

$$x_i^{(k+1)} = \frac{-1}{a_{ii}} \left(\sum_{j=1, j \neq i}^{n} a_{ij} x_j^{(k)} \right), i = 1, ..., n$$

and $H = D^{-1}(L + U)$ is called the Jacobi's method iterative matrix, where

$$h_{ij} = \begin{cases} 0, i = j \\ \frac{1}{a_{ii}}\left(l_{ij} + u_{ij}\right), i \neq j \end{cases}$$

1.4.3.2 Jacobi's Iterative Method Algorithm

To solve $AX = b$, we choose an initial approximation such as $X^{(o)} = \left(x_1^{(o)}, \ldots, x_n^{(o)}\right)$.

Step (1) $k = 1$

Step (2) For every $i = 1, \ldots, n$, we calculate:

$$x_i^{(k+1)} = \frac{-1}{a_{ii}}\left(\sum_{j=1, j\neq i}^{n} a_{ij}x_j^{(k)} - b_i\right)$$

Step (3) If $x^{(k)}$ is accurate enough, we go to the next step, otherwise we add one to k and go to step (2).

Step (4) The procedure is complete.

This algorithm requires that $a_{ii} \neq 0$ for $i = 1, \ldots, n$, otherwise the order of the equations can be changed so that a_{ii} become nonzero, and it is also better for a_{ii}'s to have the maximum possible values to accelerate convergence.

The criterion for stopping the algorithm is when

$$\|X^{(k+1)} - X^{(k)}\|$$

or

$$\frac{\|X^{(k+1)} - X^{(k)}\|}{\|X^{(k+1)}\|}$$

become less than the definite ε.

The computer program of this method is presented in the appendix number () based on the MATLAB (6.1) software.

1.4.3.3 Example

Consider the following linear system:

$$10x_1 - x_2 + 2x_3 = 6$$
$$-x_1 + 11x_2 - x_3 + 3x_4 = 25$$

$$2x_1 - x_2 + 10x_3 - x_4 = -11$$
$$3x_2 - x_3 + 4x_4 = 15$$

which has an answer of $X = (1, 2, -1, 1)'$. To obtain the iterative matrix of the Jacobi's method and the answer of the system, we proceed as follows:

$$x_1 = \frac{1}{10}x_2 - \frac{1}{5}x_3 + \frac{3}{5}$$
$$x_2 = \frac{1}{11}x_1 + \frac{1}{11}x_3 - \frac{3}{11}x_4 + \frac{25}{11}$$
$$x_3 = \frac{-1}{5}x_1 + \frac{1}{10}x_2 + \frac{1}{10}x_4 - \frac{11}{10}$$
$$x_4 = -\frac{3}{8}x_2 + \frac{1}{8}x_3 + \frac{15}{8}$$

Then

$$H = \begin{bmatrix} 0 & \frac{1}{10} & \frac{-1}{5} & 0 \\ \frac{1}{11} & 0 & \frac{1}{11} & \frac{-3}{11} \\ \frac{-1}{5} & \frac{1}{10} & 0 & \frac{1}{10} \\ & -3 & 1 & \end{bmatrix}, c = \begin{bmatrix} \frac{3}{5} \\ \frac{25}{11} \\ \frac{-11}{10} \\ \frac{15}{8} \end{bmatrix}$$

For an initial approximation, we set $X^{(0)} = (0, 0, 0, 0)'$ and produce $X^{(1)}$:

$$x_1^{(1)} = \frac{1}{10}x_2^{(0)} - \frac{1}{5}x_3^{(0)} + \frac{3}{5}$$
$$x_2^{(1)} = \frac{1}{11}x_1^{(0)} + \frac{1}{11}x_3^{(0)} - \frac{3}{11}x_4^{(0)} + \frac{25}{11}$$
$$x_3^{(1)} = \frac{-1}{5}x_1^{(0)} + \frac{1}{10}x_2^{(0)} + \frac{1}{10}x_4^{(0)} - \frac{11}{10}$$
$$x_4 = -\frac{3}{8}x_2^{(0)} + \frac{1}{8}x_3^{(0)} + \frac{15}{8}$$

So, $X^{(1)} = \left(\frac{3}{5}, \frac{25}{11}, \frac{-11}{10}, \frac{15}{8}\right)^t$ and are generated in the same way as $X^{(k)}$ until the stop condition is met. The following table shows the iteration for five steps (Table 1.1).

1.4.3.4 Theorem

If the matrix A is a strictly diagonally dominant row matrix, then the Jacobi's iterative method is convergent.

Proof The matrix A is the strictly diagonally dominant row matrix, so:

Table 1.1 Numerical results of example 1-4-3-3

k	$x_1^{(k)}$	$x_2^{(k)}$	$x_3^{(k)}$	$x_4^{(k)}$
0	0.000	0.0000	0.0000	0.0000
1	0.6000	2.3272	−0.9873	0.8789
2	1.030	2.037	−1.014	0.9844
3	1.0065	2.0036	−1.0025	0.9983
4	1.0009	2.0003	−1.0003	0.9999
5	1.0001	2.0000	−1.0000	1.0000

$$\sum_{j=1, j\neq i}^{n} |a_{ij}| < |a_{ii}|, i = 1, \ldots, n$$

and the elements of the matrix H are the iteration matrix of this method that are stated as follows:

$$h_{ij} = \begin{cases} \frac{-a_{ij}}{a_{ii}}, i \neq j \\ 0 \; i = j \end{cases}$$

Then,

$$\sum_{j=1}^{n} |h_{ij}| = \sum_{j=1, j\neq i}^{n} \left| \frac{-a_{ij}}{a_{ii}} \right| = \sum_{j=1, j\neq i}^{n} \frac{|a_{ij}|}{|a_i|} < 1$$

As a result,

$$\max_{1\leq i\leq n} \sum_{j=1}^{n} |h_{ij}| < 1$$

Therefore $\|H\|_\infty < 1$, then according to the theorem (Sect. 1.4.3.6), the iteration matrix is convergent.

We examine the convergence of the Jacobi's method according to the following theorem.

1.4.3.5 Theorem

If the matrix A is a strictly diagonally dominant row matrix, then the Jacobi's iterative method is convergent.

Proof The matrix A is the strictly diagonally dominant row matrix, so:

$$\sum_{j=1, j\neq i}^{n} |a_{ij}| < |a_{ii}|, i = 1, ..., n$$

and the elements of the matrix H are the iteration matrix of this method that are stated as follows:

$$h_{ij} = \begin{cases} \frac{-a_{ij}}{a_{ii}}, i \neq j \\ 0\ i = j \end{cases}$$

Then,

$$\sum_{j=1}^{n} |h_{ij}| = \sum_{j=1, j\neq i}^{n} \left| \frac{-a_{ij}}{a_{ii}} \right| = \sum_{j=1, j\neq i}^{n} \frac{|a_{ij}|}{|a_i|} < 1$$

As a result,

$$\max_{1\leq i\leq n} \sum_{j=1}^{n} |h_{ij}| < 1$$

Therefore $\|H\|_{\infty} < 1$, then according to the theorem (Sect. 1.4.3.4), the iteration matrix is convergent.

1.4.3.6 Gauss–Seidel Iterative Method

Now if in the system $AX = b$, the matrix A has the following decomposition:

$$A = D - L - U = (D - L) - U$$

where $U, L, D = diag(a_{11}, ..., a_{nn})$, respectively, are the lower and upper triangular matrices as follows:

$$l_{ij} = \begin{cases} -a_{ij} i > j \\ 0\ i \leq j \end{cases}, u_{ij} = \begin{cases} 0, i \geq j \\ -a_{ij} i < j \end{cases}$$

So, we have

$$[(D - L) - U]X = b$$

Then,

$$(D - L)X = UX + b$$

and as a result

$$X = (D - L)^{-1}UX + (D - L)^{-1}b$$

where $H = (D - L)^{-1}U$ is the iterative matrix of this method and to solve the system $AX = b$ we write:

$$(D - L)X^{(k+1)} = UX^{(k)} + b$$

Then,

$$DX^{(k+1)} = LX^{(k+1)} + UX^{(k)} + b$$

as a result

$$X^{(k+1)} = D^{-1}\left(LX^{(k+1)} + UX^{(k)} + b\right)$$

then

$$x_i^{(k+1)} = \frac{1}{a_{ii}}\left[b_i - \sum_{j=1}^{i-1} a_{ij}x_j^{(k+1)} - \sum_{j=i+1}^{n} a_{ij}x_j^{(k)}\right], i = 1, ..., n \tag{1.32}$$

which can be written as follows

$$\begin{aligned}
&a_{11}x_1^{(k+1)} = -a_{12}x_2^{(k)} - \cdots - a_{1n}x_n^{(k)} + b_1\\
&a_{21}x_1^{(k+1)} + a_{22}x_2^{(k+1)} = -a_{23}x_3^{(k)} - \cdots - a_{1n}x_n^{(k)} + b_2\\
&\vdots \ddots \ddots\\
&a_{n1}x_1^{(k+1)} + \cdots + a_{nn}x_n^{(k+1)} = b_n
\end{aligned}$$

1.4.3.7 Gauss–Seidel Iterative Algorithm

To solve $AX = b$, we choose an initial approximation as $X^{(0)} = \left(x_1^{(0)}, ..., x_n^{(0)}\right)$.

Step (1) $k = 1$

Step (2) For every $i = 1, ..., n$, we calculate

$$x_i^{(x+1)} = \frac{1}{a_{ii}}\left[b_i - \sum_{j=1}^{i-1} a_{ij}x_j^{(k+1)} - \sum_{j=i+1}^{n} a_{ij}x_j^{(k)}\right]$$

Step (3) If $x^{(k+1)}$ has become accurate enough, we go to the next step, otherwise, we add one to k and go to step (2).

Step (4) The procedure is complete.

This algorithm also requires that $a_{ii} \neq 0$ for every $i = 1, ..., n$, otherwise, the order of the equations can be changed so that a_{ii} becomes nonzero. Also, in this method, $x_1^{(k+1)}$ is obtained using $x_1^{(k)}$ (Jacobi's method).

The computer program of this method is presented in the appendix number (9) based on the MATLAB (6.1) software.

The convergence of the Jacobi's method usually does not result in the convergence of the Gauss–Seidel method, and vice versa, but if both methods are convergent, the Gauss–Seidel method converges faster because it uses the values $x_{i-1}^{(k+1)}, ..., x_1^{(k+1)}$ to calculate $x_i^{(k+1)}$, which are far better than $x_1^{(k)}, ..., x_{i-1}^{(k)}$.

1.4.3.8 Example

Consider the following linear system.

$$\begin{aligned}
&10x_1 - x_2 + 2x_3 = 6 \\
&-x_1 + 11x_2 - x_3 + 3x_4 = 25 \\
&2x_1 - x_2 + 10x_3 - x_4 = -11 \\
&3x_2 - x_3 + 4x_4 = 15
\end{aligned}$$

If we want to solve the system with the iterative Gauss–Seidel method

$$\begin{aligned}
x_1^{(k)} &= \frac{1}{10}x_2^{(k-1)} - \frac{1}{5}x_3^{(k-1)} + \frac{3}{5} \\
x_2^{(k)} &= \frac{1}{11}x_1^{(k)} + \frac{1}{11}x_3^{(k-1)} - \frac{3}{11}x_4^{(k-1)} + \frac{25}{11} \\
x_3^{(k)} &= \frac{-1}{5}x_1^{(k)} + \frac{1}{10}x_2^{(k)} + \frac{1}{10}x_4^{(k-1)} - \frac{11}{10} \\
x_4^{(k)} &= -\frac{3}{8}x_2^{(k)} + \frac{1}{8}x_3^{(k)} + \frac{15}{8}
\end{aligned}$$

Assuming $X^{(0)} = (0, 0, 0, 0)^t$, $X^{(1)}$ is obtained as follows:

$$X^{(1)} = (0.6, 2.3273, -0.9873, 0.8789)^t$$

In the same way, $X^{(k)}$'s are generated until the stop condition is met.

According to the $X^{(k)}$'s that are obtained, the required iterations in the Jacobi's method are almost twice as many iterations of the Gauss–Seidel method.

1.4.3.9 SOR Method

If in the system $AX = b$, we write the matrix A as follows:

$$A = (D - L) - U \tag{1.33}$$

then, we will have

$$[(D - L) - U]X = b$$

So

$$(D - L)X = UX + b$$

Now, we introduce w so that for $w = 1$, the Gauss–Seidel iterative method is obtained.

$$(D - L)X^{(k+1)} = UX^{(K)} + b$$

Therefore,

$$(D - wL)X^{(k+1)} = [(1 - w)D + wU]X^{(k)} + wb \tag{1.34}$$

Then,

$$X^{(k+1)} = (D - wL)^{-1}[(1 - w)D + wU]X^{(k)} + (D - wL)^{-1}wb \tag{1.35}$$

where

$$H_w = (D - wL)^{-1}[(1 - w)D + wU] \tag{1.36}$$

is the iterative matrix of this method.

1.4.3.10 Example

Consider the following linear system

$$\begin{aligned} &4x_1 + 3x_2 = 24 \\ &3x_1 + 4x_2 - x_3 = 30 \\ &- x_2 4x_3 = 24 \end{aligned}$$

The answer of the system is $X = (3, 4, -5)^t$. To solve the system, we use the Gauss–Seidel and SOR methods with $w = 1.25$. We suppose that $X^{(0)} = (1, 1, 1)^t$

Table 1.2 Numerical results of first seven iterations of the Gauss-Seidel method

k	$x_1^{(k)}$	$x_2^{(k)}$	$x_3^{(k)}$
0	1	1	1
1	5.250000	3.812500	−5.046875
2	3.1406250	3.8828125	−5.0292969
3	3.0878906	3.9267578	−5.0183105
4	3.0549316	3.9542236	−5.0114441
5	3.0343323	3.9713898	−5.0071526
6	3.0214577	3.9821186	−5.0044703
7	3.0134110	3.9888241	−5.0027940

is the initial vector. For every $k = 1, \ldots,$ the equations of the Gauss–Seidel method are as follows:

$$
\begin{aligned}
x_1^{(k)} &= -0.75x_2^{(k-1)} + 6 \\
x_2^{(k)} &= -0.75x_1^{(k)} + 0.25x_3^{(k-1)} + 7.5 \\
x_3^{(k)} &= 0.25x_2^{(k)} - 6
\end{aligned}
$$

and the equations of the SOR method for $w = 1.25$ are as follows:

$$
\begin{aligned}
x_1^{(k)} &= -0.25x_1^{(k-1)} - 0.9375x_2^{(k-1)} + 7.5 \\
x_2^{(k)} &= -0.9375x_1^{(k)} - 0.25x_2^{(k-1)} + 0.3125x_3^{(k-1)} + 9.375 \\
x_3^{(k)} &= 0.3125x_2^{(k)} - 0.25x_3^{(k-1)} - 7.5
\end{aligned}
$$

The first seven iterations of the Gauss–Seidel and SOR methods are given in the following tables, respectively (Tables 1.2 and 1.3).

Table 1.3 Numerical results for the first seven iterations of SOR method

k	$x_1^{(k)}$	$x_2^{(k)}$	$x_3^{(k)}$
0	1	1	1
1	6.312500	3.5195313	−6.6501465
2	2.6223145	3.9585266	−4.6004238
3	3.1333027	4.0102646	−5.0966863
4	2.9570512	4.0074838	−4.9734897
5	3.0037211	4.0029250	−5.0057135
6	2.9963276	4.0009262	−4.82822
7	3.0000498	4.0002586	−5.0003486

1.5 Numerical Integration

The antiderivatives of many functions either cannot be expressed or cannot be expressed easily in closed form (that is, in terms of known functions). Consequently, rather than evaluate definite integrals of these functions directly, we resort to various techniques of numerical integration to approximate their values. In this section, we explore three of these techniques.

1.5.1 Newton–Cotes Integration Method

These methods are generally of two types:

(1) Closed methods.
(2) Open methods.

In short, it can be said that closed methods are methods that use the two end points of the integration interval in the integration rule, and open methods are methods in which, at least one of the end points is not used. We first explain the closed methods.

One of the deterministic factors in these methods is the number of integration points used by them. For example, two-point, three-point, etc. It is well-argued that for the methods with more than or equal to eight, these methods are not cost-effective and contains the propagation of computational and rounding error. Each of the above mentioned methods is based on the choice of interpolation polynomials used instead of f, because in integration rule ($I(f)$), interpolation polynomials are used instead of f function for two reasons, first it is possible that we have not the rule of the f function and only have information about it, and second, the integration of polynomials is much easier than the integration of other functions. Now if we use a linear interpolator, the method is a two-point method, and if one of the points is not used (open Newton–Cotes), the method is one point. If we use a parabola as an interpolator, it is a three-point method and so on. Obviously, by using the interpolator instead of the f function and integrating it as an approximation, $I(f)$ will have an error because the interpolator has an error. That is, if $p_n(x)$ is the interpolator of $f(x)$ in the interval $[a, b]$ with an error of $R_n(x)$, we have:

$$f(x) = p(x) + R_n(x)$$

Then:

$$\int_a^b f(x)dx = \int_a^b p_n(x)dx + \int_a^b R_n(x)dx$$

i.e.,

$$I(f) = Q(f) + E(f), E(f) = \int_a^b R_n(x)dx$$

where it is the integration error.

Now suppose

$$p_n(x) = \sum_{i=0}^{n} L_i(x) f_i, \; L_i(x) = \prod_{j=0, j \neq i}^{n} \frac{x - x_j}{x_i - x_j}$$

where $x_j = a + jh$ and $h = \frac{b-a}{n}$. By using a change of variables as $x = a + th$ where $0 \leq t \leq n$, you can easily show that:

$$L_i(x) = \prod_{j=0, j \neq i}^{n} \frac{t - j}{i - j} = \varphi_i(t)$$

Therefore,

$$\int_a^b p_n(x)dx = h \sum_{i=0}^{n} \alpha_i f_i = \sum_{i=0}^{n} w_i f_i \tag{1.37}$$

where

$$\alpha_i = \int_0^n \varphi_i(t)dt \tag{1.38}$$

The following theorem shows that the Newton–Cotes integration weights are unique.

1.5.1.1 Theorem

Suppose $\Delta = \{a = x_0 < x_1 < \cdots < x_n = b\}$ is a fixed arbitrary partition of interval $[a, b]$. In this case, for any polynomial p with $deg(p) \leq n$, there are unique numbers of $\gamma_0, \ldots, \gamma_n$ such that:

$$\sum_{i=0}^{n} \gamma_i p(x_i) = \int_a^b p(x)dx$$

Regarding the Newton–Cotes method, the following remarks can be considered.

$$\sum_{i=0}^{n} \alpha_i = n \tag{1.39}$$

$$\alpha_i = \alpha_{n-i} \tag{1.40}$$

1.5.1.2 Remark

Suppose

$$c_i^n = \frac{1}{n} \int_0^n \varphi_i(\theta) d\theta$$

in this case

$$\sum_{i=0}^{n} c_i^n = 1 \tag{1.41}$$

$$c_i^n = c_{n-i}^n \tag{1.42}$$

The following is a concise introduction to Newton–Cotes integration methods with different numbers of points. The methods for obtaining them, as well as their error terms.

1.5.2 *The Peano's Kernel Representation*

In this section, we state and prove a theorem that can be used for easily acquiring the error of a Newton–Cotes integration rule. As previously seen in the trapezoidal and Simpson's, mean value methods, etc., no specific equation was available to obtain the error, and we could calculate the error by comparing the values and by using the Taylor expansion. But the Peano's error theorem introduces an equation for finding the error of the most Newton–Cotes integration procedures.

$$\tilde{I}(f) := \sum_{k=0}^{m_0} \alpha_{k0} f(x_{k0}) + \sum_{k=0}^{m_1} \alpha_{k1} f'(x_{k1}) + \cdots + \sum_{k=0m}^{m_n} \alpha_{kn} f(x_{kn}) \tag{1.43}$$

The integration error (1.43) can be defined as follows

$$R(f) := \tilde{I}(f) - \int_a^b f(x)dx \tag{1.44}$$

where R is a linear operator on a vector space, for example V. For instance

$$V = C^n[a, b]$$

i.e., V is a space that consists of all functions whose derivatives up to order n are continuous on $[a, b]$. Or $V = \Pi_n$ is a space consisting of all polynomials with degrees no more than n.

1.5.2.1 Theorem

Suppose that we have $R(p) = 0$ for every polynomial like $p \in \Pi_n$. That is, the integral rule is accurate for all polynomials of at most degree n.

In this case, for every function $f \in C^{n+1}[a, b]$, we have:

$$R(f) = \int_a^b f^{(n+1)}(t)k(t)dt$$

where

$$k(t) := \frac{1}{n!}R_x\left[(x-t)_+^n\right],\ (x-t)_+^n := \begin{cases} (x-t)^n, x \geq t \\ 0, x < t \end{cases}$$

and $R_x\left[(x-t)_+^n\right]$ represents the error of $(x-t)_+^n$ in the case that a function of x is considered. (The function $k(t)$ is usually called the kernel of the operator R or the Peano kernel).

Proof We write the Taylor expansion of the function $f(x)$ about the point $x = a$ as follows:

$$f(x) = \sum_{i=0}^{n} \frac{f^{(i)}(a)}{i!}(x-a)^i + r_n(x) \tag{1.45}$$

The remaining term $r_n(x)$ can be written as follows

$$r_n(x) = \frac{f^{(n+1)}(\xi)}{(n+1)!}(x-a)^{n+1} = \frac{1}{n!}\int_a^x f^{(n+1)}(t)(x-t)^n dt$$

$$= \frac{1}{n!} \int_a^x f^{(n+1)}(t)(x-t)_+^n dt$$

If we apply the operator R on (1.45), given that $R(P) = 0$ for every $p \in \Pi_n$, we have:

$$R(f) = R(r_n) = \frac{1}{n!} R_x \left(\int_a^b f^{(n+1)}(t)(x-t)_+^n dt \right) \tag{1.46}$$

Now, we prove that the operator R_x and the integral can be interchanged. To do this, first we should indicate that the differentiation and integral operators can be commuted. We show by weak induction that for $1 \leq k \leq n$,

$$\frac{d^k}{dx^k} \left[\int_a^b f^{(n+1)}(t)(x-t)_+^n dt \right] = \int_a^b f^{(n+1)}(t) \left[\frac{d^k}{dx^k} (x-t)_+^n \right] dt \tag{1.47}$$

As the initial case, we prove (1.47) for $k = 1$.

$$\begin{aligned}
\frac{d}{dx} \left[\int_a^b f^{(n+1)}(t)(x-t)_+^n dt \right] &= \frac{d}{dx} \left[\int_a^x f^{(n+1)}(t)(x-t)^n dt \right] \\
&= f^{(n+1)}(x)(x-x)^n + \int_a^x \frac{d}{dx} \left[f^{(n+1)}(t)(x-t)^n \right] dt \\
&= \int_a^x f^{(n+1)}(t) \frac{d}{dx} \left[(x-t)^n \right] dt
\end{aligned}$$

As the induction hypothesis, we assume that Eq. (1.47) holds for $k = n - 1$.

$$\begin{aligned}
\frac{d^{n-1}}{dx^{n-1}} \left[\int_a^b f^{(n+1)}(t)(x-t)_+^n dt \right] &= \int_a^b f^{(n+1)}(t) \frac{d^{n-1}}{dx^{n-1}} \left[(x-t)_+^n \right] dt \\
&= n! \int_a^b f^{(n+1)}(t)(x-t)_+ dt \\
&= n! \int_a^x f^{(n+1)}(t)(x-t)_+ dt
\end{aligned}$$

We prove the induction step for $k = n$.

$$\frac{d^n}{dx^n}\left[\int_a^b f^{(n+1)}(t)(x-t)_+^n dt\right] = \frac{d}{dx}\left[\frac{d^{n-1}}{dx^{n-1}}\int_a^b f^{(n+1)}(t)(x-t)_+^n dt\right]$$
$$= n!\frac{d}{dx}\left[\int_a^b f^{(n+1)}(t)(x-t)_+ dt\right]$$
$$= \int_a^x f^{(n+1)}(t) \times n! dt$$
$$= \int_a^x f^{(n+1)}(t) \times \frac{d^n}{dx^n}(x-t)^n dt$$

i.e.,

$$\frac{d^n}{dx^n}\left[\int_a^b f^{(n+1)}(t)(x-t)_+^n dt\right] = \int_a^x f^{(n+1)}(t)\left[\frac{d^n}{dx^n}(x-t)_+^n\right]dt$$

it can be said that the operator $\tilde{I}$ can also be commutated by the integral.

Now, to indicate that the operator R_x can be commutated by the integral, it is sufficient that by considering Eq. (1.47) we show that:

$$\int_a^b\left[\int_a^b f^{(n+1)}(t)(x-t)_+^n dt\right]dx = \int_a^x f^{(n+1)}(t)\left[\int_a^b (x-t)_+^n dx\right]dt$$

The above equation is always established according to the continuity properties of the following function:

$$f^{(n+1)}(t)(x-t)_+^n$$

So it can be claimed as follows that R_x is commutated by the integral.

$$\int_a^b f^{(n+1)}(t)R_x\left[(x-t)_+^n\right]dt = \int_a^b f^{(n+1)}(t)\left[\tilde{I}\left[(x-t)_+^n\right] - \int_a^b (x-t)_+^n dx\right]dt$$
$$= \int_a^b f^{(n+1)}\tilde{I}\left[(x-t)_+^n\right]dt - \int_a^b\left[\int_a^b f^{(n+1)}(t)(x-t)_+^n dx\right]dt$$

$$= \tilde{I}\left[\int_a^b f^{(n+1)}(x-t)_+^n dt\right] - \int_a^b\left[\int_a^b f^{(n+1)}(t)(x-t)_+^n dt\right]dx$$

$$= R_x\left[\int_a^b f^{(n+1)}(t)(x-t)_+^n dt\right]dx$$

So, we can write Eq. (1.46) as follows

$$R(f) = \frac{1}{n!}\int_a^b f^{(n+1)}(t)R_x\left[(x-t)_+^n\right]dt$$

$$R(f) = \int_a^b f^{(n+1)}(t)\frac{1}{n!}R_x\left[(x-t)_+^n\right]dt$$

$$:= \int_a^b f^{(n+1)}(t)k(t)dt$$

where

$$k(t) := \frac{1}{n!}R_x\left[(x-t)_+^n\right]$$

Given the continuity of $f^{(n+1)}$ on $[a, b]$ and supposing that $k(t)$ does not change the sign in the same interval, according to the integral mean value theorem we can write:

$$R(f) = f^{(n+1)}(\xi)\int_a^b k(t)dt, \exists\xi \in (a, b) \tag{1.48}$$

By considering Eq. (1.46) and comparing it with (1.48) and given that for every

$$P \in \Pi_n, R(P) = 0$$

we have

$$\begin{aligned} R(f) &= \frac{f^{(n+1)}(\xi)}{(n+1)!}R\left[(x-a)^{n+1}\right] \\ &= \frac{f^{(n+1)}(\xi)}{(n+1)!}R\left(x^{n+1}\right), \exists\xi \in (a, b) \end{aligned} \tag{1.49}$$

the fixed error coefficient can be calculated, i.e.,

$$\frac{R\left(x^{n+1}\right)}{(n+1)!}$$

1.5.3 Gauss Integration Method

If we want to discuss more straightforward, we should write an equation for $m = 0$ as follows:

$$\int_a^b f(x)dx = \sum_{j=1}^{n} H_j f\left(a_j\right) + E \quad (1.50)$$

if we have

$$\left(f(x) = x^k \& k = 0, 1, ..., 2n - 1\right) \Rightarrow E = 0$$

in this case, we have a system of nonlinear equations of order $2n$ that is quadratic and has $2n$ unknowns. According to the above-mentioned discussion, the right side of (1.50) can be written as follows:

$$\alpha_k = \sum_{j=0}^{n} H_j a_j^k, k = 0, 1, ..., 2n - 1 \quad (1.51)$$

If we equate it to the left we have

$$\alpha_k = \int_a^b x^k dx = \frac{1}{k+1}\left(b^{k+1} - a^{k+1}\right) \quad (1.52)$$

The system of nonlinear equations is written as follows

$$\sum_{j=0}^{n} H_j a_j^k = \frac{1}{k+1}\left(b^{k+1} - a^{k+1}\right), k = 0, 1, ..., 2n - 1 \quad (1.53)$$

In fact, by solving the system (1.53), we can obtain the unknowns in such a way that the integration rule (1.50) is exact for the polynomial functions of degree at most $2n - 1$.

Given the nonlinearity of (1.53), it is not easy to solve it for large orders (large n). But it is more accurate than Newton–Cotes methods. As we have already seen, in the Newton–Cotes method, only unknowns were the integral weights, whereas in these methods, in addition to the integration weights, a_j's are unknown, too. So, we have $2n$ unknowns and need $2n$ equations. It should be noted that a_j's are the roots of orthogonal polynomials. In this section, we deal with orthogonal polynomials that are briefly discussed in chapter three. Considering that orthogonal polynomials are defined in a particular interval and by their specific weight function, so there are corresponding Gaussian numerical integration methods for each set of orthogonal polynomial functions and their roots. First, without the consideration of the weight function and given that the error term of such methods must be proportional to the $2n$th derivative of the function f, we obtain the general form of Gaussian method using simple Hermite interpolation for n points of $a_1, a_2, \ldots, a_n$. If we write the simple Hermite interpolation formula for the mentioned n points with the new notation, we have:

$$f(x) = \sum_{j=1}^{n} h_j(x) f(a_j) + \sum_{j=1}^{n} \bar{h}_j(x) f'(a_j) + \frac{R^2(x)}{(2n)!} f^{(2n)}(\xi) \tag{1.54}$$

where

$$R(x) = \prod_{j=1}^{n} (x - a_j) \tag{1.55}$$

$$h_j(x) = [1 - 2l'_j(a_j)(x - a_j)] l_j^2(x), \bar{h}_j(x) = (x - a_j) l_j^2(x) \tag{1.56}$$

It is obvious that the error term of Eq. (1.54) for polynomial functions of at most degree $2n - 1$ is equal to zero. By integrating both sides of Eq. (1.54) on the interval $[a, b]$, the following equation can be obtained.

$$\int_a^b f(x)dx = \sum_{j=1}^{n} H_j f(a_j) + \sum_{j=1}^{n} \bar{H}_j f'(a_j) + E \tag{1.57}$$

where

$$H_j = \int_a^b h_j(x)dx, \bar{H}_j = \int_a^b \bar{h}_j(x)dx \tag{1.58}$$

are integration weights and their error are:

$$E = \int_a^b \frac{R^2(x)}{(2n)!} f^{(2n)}(\xi)dx \tag{1.59}$$

As it was mentioned before, if f is a polynomial of at most degree $2n-1$, then $E=0$. Now if we choose a_j's so that $\bar{H}_j=0$ for every j, then Eq. (1.57) reduced to Eq. (1.50) and satisfy its properties. So, it can be said that $\bar{H}_j=0$ is a sufficient as well as necessary condition to achieve the accuracy of $2n-1$. Therefore, the following lemma can be expressed and proved.

1.5.3.1 Lemma

If (1.50) is accurate for the polynomials of at most degree $2n-1$, then the necessary and sufficient condition for Eq. (1.57) to be accurate for the polynomials of at most degree $2n-1$ is that $\bar{H}_j=0$ for every j.

Proof First, we prove the necessity. Suppose in Eq. (1.57) we have:

$$\left(f(x)=x^k \& k=0,1,\ldots,2n-1\right) \Rightarrow E=0$$

That is, for every polynomial of at most degree $2n-1$, the error term is equal to zero. So, if we choose $f(x)$ as $\bar{h}_k(x)$, then $E=0$. Because $\bar{h}_k$ is a polynomial of at most degree $2n-1$ (why?). So, we have:

$$\int_a^b \bar{h}_k(x)dx=\sum_{j=1}^{n} H_j\bar{h}_k(a_j)+\sum_{j=1}^{n} \bar{H}_j\bar{h}'_k(a_j),\ k=1,2,\ldots,n \tag{1.60}$$

According to the properties of the polynomials h_k, $\bar{h}_k$ and Eq. (1.58) we have:

$$\int_a^b \bar{h}_k(x)dx=\bar{H}_k$$

Because

$$\bar{h}_k(a_j)=0,\ \bar{h}'_k(a_j)=\delta_{jk}$$

On the other hand, according to (1.49) and given that

$$f(x)=\bar{h}_k(x)$$

so

$$\int_a^b \bar{h}_k(x)dx=\sum_{j=1}^{n} H_j\bar{h}_k(a_j)$$

That is, for every k, $\bar{H}_k = 0$. Therefore, the sentence is true. The proof of adequacy is straightforward.

According to the lemma (Sect. 1.5.3.1), we can show that under the following conditions

$$\bar{H}_j = \int_a^b \bar{h}_j(x)dx = 0$$

i.e.

$$\int_a^b (x - a_j)l_j^2(x)dx = \int_a^b R(x) \cdot \frac{l_j(x)}{R'(a_j)}dx = 0 \tag{1.61}$$

Since $R(x)$ is of degree n and $l_j(x)$ is of degree $n-1$, then it should be claimed that the necessary and sufficient condition for $\bar{H}_j = 0$ is that $R(x)$ is perpendicular to all the polynomials of at most degree $n-1$ on $[a, b]$.

If we want to argue the set of the Legendre orthogonal polynomials, since the coefficient of the leading term of $R(x)$ is equal to 1, then

$$R(x) = \frac{2^n(n!)^2}{(2n)!}P_n(x) \tag{1.62}$$

It was observed that the Legendre polynomials, having a constant weight of one, form an orthogonal set on the interval $[-1, 1]$. For this purpose, it should be noted that any desired interval $[a, b]$ can be converted to the interval $[-1, 1]$ by a change of variables as $y = \frac{2x-a-b}{b-a}$.

In the following table, the zeros of the Legendre polynomial up to degree 5 along with their weights are listed.

Now, we calculate the Gauss–Legendre integration weight.

$$\begin{aligned}
H_j &= \int_{-1}^{1} h_j(x)dx = \int_{-1}^{1} [1 - 2l'_j(a_j)(x - a_j)]l_j^2(x)dx \\
&= \int_{-1}^{1} l_j^2(x)dx - 2l'_j \int_{-1}^{1} (x - a_j)l_j^2(x)dx \\
&= \int_{-1}^{1} l_j^2(x)dx - 2l'_j(a_j)\bar{H}_j
\end{aligned}$$

$$= \int_{-1}^{1} l_j^2(x)dx \tag{1.63}$$

The above equation indicates that all of the Gauss–Legendre integration weights are positive. If we want to have a simpler form for calculating the weights of H_j, we should employ Eq. (1.50) such that in this equation, the term E for the function $f(x) = l_k(x)$ become zero. (Why?).

Considering $l_k(a_j) = \delta_{kj}$, we have:

$$\int_{-1}^{1} l_k(x)dx = \sum_{j=1}^{n} H_j l_k(a_j) = H_k \tag{1.64}$$

By comparing (1.63) and (1.64), we can write:

$$\int_{-1}^{1} l_j^2(x)dx = \int_{-1}^{1} l_j(x)dx \tag{1.65}$$

The error term (1.59) in this method is expressed as

$$E = \frac{f^{(2n)}(\eta)}{(2n)!} \int_{-1}^{1} R^2(x)dx \tag{1.66}$$

where $\eta \in (-1, 1)$. Of course, it should be noted that since $R^2(x)$ does not change the sign on $[-1, 1]$, the integral mean value theorem has been used.

Therefore, the Gauss–Legendre integration method was introduced where the integration weights are obtained from (1.64) and the integration points are the roots of the Legendre polynomials.

1.5.3.2 Example

We compute $\int_1^3 \frac{dx}{x}$ using the three-point Gauss–Legendre method.

Solution: If we use the corresponding change of variables, the transformation $y = x - 2$ converts the interval $[1, 3]$ to the interval $[-1, 1]$. Using Table 1.4 for $n = 3$ we have:

$$\int_{-1}^{1} \frac{dy}{y+2} \simeq \frac{5}{9} \times \frac{1}{1.225403} + \frac{8}{9} \times \frac{1}{2} + \frac{5}{9} \cdot \frac{1}{2.774597} \simeq 1.098039$$

Table 1.4 Zeros of the Legendre polynomial up to degree 5 with their weights

n	a_j	H_j
2	$\pm 0.577350 = \pm \frac{1}{\sqrt{3}}$	1
3	0	$\frac{8}{9}$
	± 0.774597	$\frac{5}{9}$
4	± 0.339981	0.652145
	± 0.861136	0.347855
5	0	0.568889
	± 0.538469	0.478629
	± 0.906180	0.236927

Using (1.66), we have:

$$E = \frac{f^{(6)}(\eta)}{6!} \int_{-1}^{1} R^2(x)dx$$

where

$$R(x) = x(x - 0.774597)(x + 0.774597)$$

Chapter 2
Interval Interpolation

Sometimes the points used for interpolation may not be exactly available, and we may have some parts of a data. For this type of data, the word "approximately" is commonly used. Now, if we want to make the word approximately and about meaningful, one of the ways is to use the interval; for example, we show about 2 with $[2-\varepsilon, 2+\varepsilon]$, where ε is introduced arbitrarily. So instead of about 2 or approximately 2, you can use a range that contains 2.

2.1 Interval Error

In this section, in order to discuss interval data error, we will first have a brief overview on the interval calculations.

2.1.1 Interval Calculations

If machine numbers are an approximation of numbers, i.e.,

$$a \in \mathbb{R}\ominus, \mathrm{a} \in \mathbb{Q}$$

where

$$\tilde{a} := [\underline{a}, \overline{a}] = \{x | x \in \mathbb{R}, \underline{\mathrm{a}} \leq \mathrm{x} \leq \overline{\mathrm{a}}\}$$

In this case, we say $\tilde{a}$ is an interval number.

Arithmetic operations on interval numbers are defined as follows.

(1) The sum of two intervals

T. Allahviranloo and A. Esfandiari, *A Course on Integral Equations with Numerical Analysis*, Mathematical Engineering, https://doi.org/10.1007/978-3-030-85350-1_2

$$\tilde{a} + \tilde{b} = [\underline{a}, \overline{a}] + [\underline{b}, \overline{b}] = [\underline{a} + \overline{b}, \overline{a} + \overline{b}]$$

(2) Symmetry of an interval number

$$-\tilde{a} = -[\underline{a}, \overline{a}] = [-\overline{a}, -\underline{a}]$$

(3) The difference between two interval numbers:

$$\tilde{a} - \tilde{b} = [\underline{a}, \overline{a}] - [\underline{b}, \overline{b}] = [\underline{a} - \overline{b}, \overline{a} - \underline{b}]$$

(4) The multiplication of a scalar to an interval number:

$$k\tilde{a} = k[\underline{a}, \overline{a}] = \begin{cases} [k\underline{a}, k\overline{a}], k \geq 0 \\ [k\overline{a}, k\underline{a}], k < 0 \end{cases}$$

(5) The product of two intervals

$$\tilde{a} \cdot \tilde{b} = [\underline{a}, \overline{a}] \cdot [\underline{b}, \overline{b}] = [\underline{c}, \overline{c}]$$

where

$$\underline{c} = \min\{\underline{a} \cdot \underline{b}, \underline{a} \cdot \overline{b}, \overline{a} \cdot \underline{b}, \overline{a} \cdot \overline{b}\}$$
$$\overline{c} = \max\{\underline{a} \cdot \underline{b}, \underline{a} \cdot \overline{b}, \overline{a} \cdot \underline{b}, \overline{a} \cdot \overline{b}\}$$

(6) The division of two interval number:

$$\frac{\tilde{a}}{\tilde{b}} = \frac{[\underline{a}, \overline{a}]}{[\underline{b}, \overline{b}]} = [\underline{c}, \overline{c}]$$

where

$$\underline{c} = \min\left\{\frac{\underline{a}}{\underline{b}}, \frac{\overline{a}}{\underline{b}}, \frac{\underline{a}}{\overline{b}}, \frac{\overline{a}}{\overline{b}}\right\}$$
$$\overline{c} = \max\left\{\frac{\underline{a}}{\underline{b}}, \frac{\overline{a}}{\underline{b}}, \frac{\underline{a}}{\overline{b}}, \frac{\overline{a}}{\overline{b}}\right\}$$

2.2 Interval Interpolation

Classification of interval data is possible in two ways.

1. Data whose length and width are interval.

2. Data that only whose width is an interval.

Checking case (1) is more difficult than checking case (2). Because in case (1), the resulting interpolation criterion will have a complex and problematic form, so we will ignore it and only examine case (2). In this case, it is assumed that $n + 1$ points are as follows:

$$\left(x_i, \left[\underline{f}_i, \overline{f}_i\right]\right), i = 0, \ldots, n$$

If we want to consider the interpolation problem

$$\phi(x; a_0, \ldots, a_n) = \sum_{j=0}^{n} a_j \phi_j(x) \tag{2.1}$$

Such that the interpolation condition is satisfied, we have the following interval data

$$\phi(x; a_0, \ldots, a_n) = \left[\underline{f}_i, \overline{f}_i\right], i = 0, \ldots, n$$

According to the right side of the equation, it is obvious that the left side must also be an interval. Therefore, it can be written

$$\left[\underline{\phi}_i, \overline{\phi}_i\right] = \left[\underline{f}_i, \overline{f}_i\right], i = 0, \ldots, n$$

where

$$\underline{\phi} = \min\left\{u | \underline{\phi} \le u \le \overline{\phi}\right\}$$
$$\overline{\phi} = \max\left\{u | \underline{\phi} \le u \le \overline{\phi}\right\}$$

So, according to the equality between two intervals, we have:

$$L : \underline{\phi}(x_i; a_0, \ldots, a_n) = f_i$$
$$U : \overline{\phi}(x_i; a_0, \ldots, a_n) = \overline{f}_i, i = 0, \ldots, n$$

So for every i, we have two interpolation problems of L and U. So totally, we will have two system of order $n + 1$. According to Eq. (2.1), $\underline{\phi}$ can be introduced as follows:

$$\underline{\phi}(x; a_0, \ldots, a_n) = \underline{\sum_{j=0}^{n} a_j \phi_j(x)} = \sum_{j=0}^{n} \underline{a_j \phi_j(x)}$$

By multiplying a scalar to the interval, it can be claimed that:

$$\underline{a_j\phi_j(x)} = \begin{cases} \underline{a}_j\phi_j(x), \phi_j(x) \geq 0 \\ \overline{a}_j\phi_j(x), \phi_j(x) < 0 \end{cases}$$

Therefore, it can be written

$$\underline{\phi}(x; a_0, ..., a_n) = \sum_{\phi_j(x)\geq 0} \underline{a}_j\phi_j(x) + \sum_{\phi_j(x)<0} \overline{a}_j\phi_j(x) \tag{2.2}$$

So interpolation problem (L) is obtained as follows.

$$\underline{\phi}(x_i; a_0, ..., a_n) = \sum_{\phi_j(x_i)\geq 0} \underline{a}_j\phi_j(x_i) + \sum_{\phi_j(x_i)<0} \overline{a}_j\phi_j(x_i)$$

$$= \underline{f}_i, i = 0, ..., n$$

It can also be said that

$$\overline{\phi}(x; a_0, ..., a_n) = \overline{\sum_{j=0}^{n} a_j\phi_j(x)} = \sum_{j=0}^{n} \overline{a_j\phi_j(x)}$$

and as before

$$\overline{a_j\phi_j(x)} = \begin{cases} \overline{a}_j\phi_j(x), \phi_j(x) \geq 0 \\ \underline{a}_j\phi_j(x), \phi_j(x) < 0 \end{cases}$$

So we will have

$$\overline{\phi}(x; a_0, ..., a_n) = \sum_{\phi_j(x)\geq 0} \overline{a}_j\phi_j(x) + \sum_{\phi_j(x)<0} \underline{a}_j\phi_j(x) \tag{2.3}$$

and the interpolation problem (U) is as follows

$$\overline{\phi}(x_i; a_0, ..., a_n) = \sum_{\phi_j(x_i)\geq 0} \overline{a}_j\phi_j(x_i) + \sum_{\phi_j(x_i)<0} \underline{a}_j\phi_j(x_i)$$

$$= \overline{f}_i, i = 0, ..., n$$

as a result

$$
\begin{aligned}
L &: \sum_{\phi_j(x_i)\geq 0} \underline{a}_j\phi_j(x_i) + \sum_{\phi_j(x_i)<0} \overline{a}_j\phi_j(x_i) = \underline{f}_i \\
U &: \sum_{\phi_j(x_i)\geq 0} \overline{a}_j\phi_j(x_i) + \sum_{\phi_j(x_i)<0} \underline{a}_j\phi_j(x_i) = \underline{f}_i \\
i &= 0
\end{aligned}
\tag{2.4}
$$

According to Eqs. (2.2) and (2.3), it can be said that $\overline{\phi}$ and $\underline{\phi}$ form an interval. So, in this case, we always have two types of interpolation problems as (L) and (U), each of which must uniquely exist, and (L) and (U) interpolate points $\left(x_i, \underline{f}_i\right)$ and $\left(x_i, \overline{f}_i\right)$, respectively. So, in this case, instead of an interpolating function, we have a band of an interpolating function (Fig. 2.1).

Therefore, it is necessary for each group of the points $\left(x_i, \underline{f}_i\right)$ and $\left(x_i, \overline{f}_i\right)$, $i = 0, ..., n$ to form a function, and an interpolation problem corresponded to each of them is introduced.

Now if $\phi_j(x) = x^j$, the interpolation interval is as follows

$$
\begin{aligned}
L &: \underline{p}(x) = \sum_{x^j\geq 0} \underline{a}_j x^j + \sum_{x^j<0} \overline{a}_j x^j \\
U &: \overline{p}(x) = \sum_{x^j\geq 0} \overline{a}_j x^j(x_i) + \sum_{x^j<0} \underline{a}_j x^j(x_i)
\end{aligned}
\tag{2.5}
$$

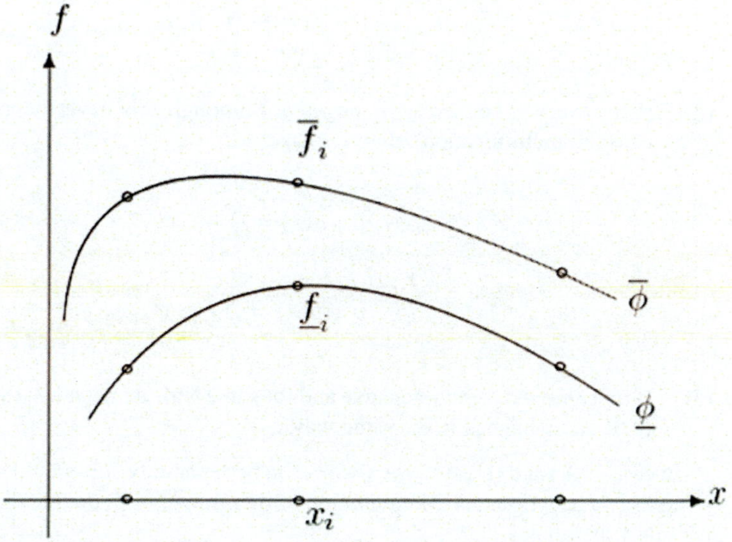

Fig. 2.1 Interval function

It is clear that by considering the j th power of x, as even and odd, we can write:

$$L : \underline{p}(x) = \sum_{j=2k} \underline{a}_j x^j + \sum_{j=2k+1} \overline{a}_j x^j x < 0$$

$$\underline{p}(x) = \sum_{j=0}^{n} \underline{a}_j x^j x \geq 0$$

$$k = 0, \ldots, \left[\frac{n}{2}\right]$$

$$U : \overline{p}(x) = \sum_{j=2k} \overline{a}_j x^j + \sum_{j=2k+1} \underline{a}_j x^j x < 0$$

$$\overline{p}(x) = \sum_{j=0}^{n} \overline{a}_j x^j x \geq 0$$

$$k = 0, \ldots, \left[\frac{n}{2}\right]$$

The following cases can be considered for system (2.5):

(1) If for every x and every j, x^j is non-negative, then

$$\begin{aligned} L : \underline{p}(x) &= \sum_{j=0}^{n} \underline{a}_j x^j \\ U : \overline{p}(x) &= \sum_{j=0}^{n} \overline{a}_j x^j \end{aligned} \tag{2.6}$$

(2) If for every x and every, is negative (although this never happens due to a constant multiplication), then we have

$$\begin{aligned} L : \underline{p}(x) &= \sum_{j=0}^{n} \overline{a}_j x^j \\ U : \overline{p}(x) &= \sum_{j=0}^{n} \underline{a}_j x^j \end{aligned} \tag{2.7}$$

(3) If for some and, is non-negative and for some and, are negative, then the system (2.5) is established in the same way.

In case 1, (L) and (U) systems always can be separately solved and also these two systems can be written as a combination of themselves with double dimensions and examines as follows

$$\begin{bmatrix} B & 0 \\ 0 & B \end{bmatrix} \begin{bmatrix} \underline{a} \\ \overline{a} \end{bmatrix} = \begin{bmatrix} \underline{f} \\ \overline{f} \end{bmatrix}$$

where

$$B = \left[x_i^j\right]_{i,j=0}^n$$
$$\underline{a} = \left[\underline{a}_0, ..., \underline{a}_n\right]^t, \overline{a} = [\overline{a}_0, ..., \overline{a}_n]$$
$$\underline{f} = \left[\underline{f}_0, ..., \underline{f}_n\right]^t, \overline{f} = [\overline{f}_0, ..., \overline{f}_n]^t$$

In case (2), (L) and (U) systems always can still be separately solved. But with double dimensions, the combination is as follows:

$$\begin{bmatrix} 0 & C \\ C & 0 \end{bmatrix} \begin{bmatrix} \underline{a} \\ \overline{a} \end{bmatrix} = \begin{bmatrix} \underline{f} \\ \overline{f} \end{bmatrix}$$

where $\underline{a}, \overline{a}, \underline{f}$, and $\overline{f}$ are the same as the above vectors, and matrix C is defined similar to matrix B. In case (3), if we want to solve systems (L) and (U) separately, then we either have no answer or have infinite answer, because, the number of unknowns exceeds the number of constraints. So, we have to solve the combined system of double order, which is obtained as follows:

$$\begin{bmatrix} B & C \\ C & B \end{bmatrix} \begin{bmatrix} \underline{a} \\ \overline{a} \end{bmatrix} = \begin{bmatrix} \underline{f} \\ \overline{f} \end{bmatrix} \tag{2.8}$$

where $B \geq 0$ and $C \leq 0$.

In all three cases of (1), (2) and (3), it can be claimed that the matrix form is as (2.6). So assuming

$$S = \begin{bmatrix} B & C \\ C & B \end{bmatrix}, X = \begin{bmatrix} \underline{a} \\ \overline{a} \end{bmatrix}, Y = \begin{bmatrix} \underline{f} \\ \overline{f} \end{bmatrix}$$

For convenience, we consider the system $SX = Y$. Now, we have to try to solve and discuss the existence and uniqueness of the answer and also the relationship between its answer and the answer of the interpolation problem

$$\sum_{j=0}^{n} a_j x_i^j = \left[\underline{f}_i, \overline{f}_i\right], i = 0, ..., n$$

In this case

$$A = \begin{bmatrix} 1 & x_0 & \cdots & x_0^n \\ 1 & x_1 & \cdots & x_1^n \\ \vdots & \vdots & & \vdots \\ 1 & x_n & \cdots & x_n^n \end{bmatrix} = B + C$$

And by assuming that for every i, points x_i are distinct, then $\det(B + C) \neq 0$.

2.2.1 *Theorem*

$$\det(S) = \det(B - C) \cdot \det(B + C)$$

Proof If we subtract the last $n + 1$ rows from the first $n + 1$ rows, respectively, we have

$$\det(S) = \det\begin{bmatrix} B - C & C - B \\ C & B \end{bmatrix}$$

And now if we add the first $n+1$ columns to the second $n+1$ columns, respectively, we have

$$\det(S) = \det\begin{bmatrix} B - C & 0 \\ C & B + C \end{bmatrix}$$

where according to the Laplace expansion of determinant with respect to the first $n + 1$ columns, the sentence is true. As it seen, the interpolation with these interval data leads to solve the linear block systems. To solve these block systems the block Jacobi and block Gauss iterative methods can be used. For this purpose, these methods are explained as follow.

First, we are going to proof the following theorems.

2.2.2 *Theorem*

Let S be nonsingular. Then the unique solution X of $SX = Y \rightarrow \begin{bmatrix} S_1 \geq 0 & S_2 \leq 0 \\ S_2 \leq 0 & S_1 \geq 0 \end{bmatrix}$ is always an interval vector for arbitrary vector Y, if S^{-1} is nonnegative.

Proof It is clear that S^{-1} have the same structure like S, i.e.

$$S^{-1} = \begin{bmatrix} B & C \\ C & B \end{bmatrix}$$

From $X = S^{-1}Y$, we have

$$\begin{cases} B\underline{Y} - C\overline{Y} = \underline{X} \\ -C\underline{Y} - B\overline{Y} = \overline{X} \end{cases} \tag{2.9}$$

Thus

$$(\overline{X} - \underline{X}) = (B + C)(\overline{Y} - \underline{Y}) \geq 0 \tag{2.10}$$

Because $(S^{-1})_{ij} = t_{ij} \geq 0$ and $(\overline{Y} - \underline{Y}) \geq 0$. Since $\overline{Y}$ is monotonically decreasing and $\underline{Y}$ is monotonically increasing, Eq. (2.9) is also necessary and sufficient for $\overline{X}$ and $\underline{X}$ to be monotonically decreasing and increasing, respectively. The bounded left continuity of $\underline{X}$ and $\overline{X}$ is obvious since they are linear combinations of $\underline{Y}$ and $\overline{Y}$. The proof in completed.

2.2.3 Theorem

The matrix A in below is strictly diagonally dominant if and only if matrix S be strictly diagonally dominant.

$$\begin{gathered} a_{11}x_1 + a_{13}x_2 + \ldots + a_{1n}x_n = y_1, \\ a_{21}x_1 + a_{23}x_2 + \ldots + a_{2n}x_n = y_2, \\ \vdots \\ a_{n1}x_1 + a_{n3}x_2 + \ldots + a_{nn}x_n = y_n \end{gathered}$$

Proof Let A be column strictly diagonally dominant, i.e. $|a_{jj}| > \sum_{i=1,i\neq j}^{n} |a_{ij}|$, $j = 1, \ldots, n$. By considering the structure of S we have

$$\begin{aligned} s_{ij} = s_{n+i,n+j} = a_{aj} > 0 \leftrightarrow s_{n+i,j} = s_{i,n+j} = a_{ij} = 0 \\ s_{ij} = s_{n+i,n+j} = a_{aj} = 0 \leftrightarrow s_{n+i,j} = s_{i,n+j} = a_{ij} < 0 \end{aligned} \tag{2.11}$$

also

$$\sum_{i=1,i\neq j}^{2n} |s_{ij}| = \sum_{i=1,j\neq j}^{n} |s_{ij}| + \sum_{i=1,i\neq j}^{n} |s_{n+ij}|,\ j = 1, 2, \ldots, 2n$$

If $s_{ij} > 0$, $i, j = 1, 2, \ldots, n$ then

$$\sum_{i=1,i\neq j}^{n} |s_{ij}| = \begin{cases} \sum_{i=1,i\neq j}^{2n} |s_{ij}| \\ \sum_{i=1,i\neq j}^{2n} |s_{i,n+j}| \end{cases}$$

$$= \begin{cases} \sum_{i=1,i\neq j}^{n} |s_{ij}| + \sum_{i=1,i\neq j}^{n} |s_{n+i,j}| \\ \sum_{i=1,i\neq j}^{n} |s_{i,n+j}| + \sum_{i=1,i\neq j}^{n} |s_{n+i,n+j}|, \; j = 1, 2, \ldots, n \end{cases}$$

From (2.11)

$$\sum_{i=1,i\neq j}^{2n} |s_{ij}| \begin{cases} \sum_{i=1,i\neq j}^{n} |s_{ij}| < |a_{jj}| = |s_{jj}| \\ \sum_{i=1,i\neq j}^{n} |s_{ij}| < |a_{jj}| = |s_{n+i,n+j}|, \; j = 1, 2, \ldots, n \end{cases}$$

Then

$$\sum_{i=1,i\neq j}^{2n} |s_{ij}| < |s_{jj}|, \; j = 1, 2, \ldots, n$$

Now suppose that S be column strictly diagonally dominant, we have

$$\sum_{i=1,i\neq j}^{2n} |s_{ij}| = \sum_{i=1,i\neq j}^{2n} |s_{ij}| + \sum_{i=1,i\neq j}^{2n} |s_{n+i,j}|$$

By considering (2.11) and $A = S_1 + S_2$ we have

$$\sum_{i=1,i\neq j}^{2n} |s_{ij}| = \sum_{i=1,i\neq j}^{2n} |a_{ij}| < |s_{jj}| = |a_{jj}|, \; j = 1, \ldots, n$$

Since $s_{ij} = 0$, $j = n, \ldots, 2n$. It can be hold for now strictly diagonally dominant too.

Without loss of generality, suppose that $s_{ii} > 0$ for all $i = 1, \ldots, 2n$. Let $S = D + L + U$ where

$$D = \begin{bmatrix} D_1 & 0 \\ 0 & D_1 \end{bmatrix}, L = \begin{bmatrix} L_1 & 0 \\ 0 & L_1 \end{bmatrix}, U = \begin{bmatrix} U_1 & S_2 \\ 0 & U_1 \end{bmatrix}$$

$(D_1)_{ii} = s_{ii} > 0, i = 1, \ldots, n$ and suppose$S_1 = D_1 + L_1 + U_1$. In the Jacobi method, from the structure of $SX = Y$ we have

$$\begin{bmatrix} D_1 & 0 \\ 0 & D_1 \end{bmatrix}\begin{bmatrix} \underline{X} \\ \overline{X} \end{bmatrix} + \begin{bmatrix} L_1 + U_1 & S_2 \\ S_2 & L_1 + U_2 \end{bmatrix}\begin{bmatrix} \underline{X} \\ \overline{X} \end{bmatrix} = \begin{bmatrix} \underline{Y} \\ \overline{Y} \end{bmatrix}$$

Then

$$\begin{aligned} \underline{X} &= D_1^{-1}\underline{Y} - D_1^{-1}(L_1 + U_1)\underline{X} - D_1^{-1}S_2\overline{X} \\ \overline{X} &= D_1^{-1}\overline{Y} - D_1^{-1}(L_1 + U_1)\overline{X} - D_1^{-1}S_2\underline{X} \end{aligned} \tag{2.12}$$

So, the Jacobi iterative technique will be

$$\begin{aligned} \underline{X}^{k+1} &= D_1^{-1}\underline{Y} - D_1^{-1}(L_1 + U_1)\underline{X}^k - D_1^{-1}S_2\overline{X}^k \\ \overline{X}^{k+1} &= D_1^{-1}\overline{Y} - D_1^{-1}(L_1 + U_1)\underline{X}^k - D_1^{-1}S_2\underline{X}^k,\ k = 0, 1, \ldots \end{aligned} \tag{2.13}$$

The elements of $X^{k+1} = \left(\underline{X}^{k+1}, \overline{X}^{k+1}\right)^t$ are

$$\underline{x}_i^{k+1} = \frac{1}{s_{i,i}} = \left[\underline{y}_i - \sum_{j=1, j\neq i}^{n} s_{i,j}\underline{x}_j^k - \sum_{j=1}^{n} s_{i,n+j}\overline{x}_j^k\right]$$

$$\overline{x}_j^{k+1} = \frac{1}{s_{i,i}} = \left[\overline{y}_i - \sum_{j=1, j\neq i}^{n} s_{i,j}\overline{x}_j^k - \sum_{j=1}^{n} s_{i,n+j}\underline{x}_j^k\right]$$

The results in the matrix form of the Jacobi iterative technique are $X^{k+1} = PX^k + C$ where

$$= \begin{bmatrix} -D_1^{-1}(L_1 + U_1) & -D_1^{-1}S_2 \\ -D_1^{-1}S_2 & -D_1^{-1}(L_1 + U_1) \end{bmatrix}, C = \begin{bmatrix} D_1^{-1}\underline{Y} \\ D_1^{-1}\overline{Y} \end{bmatrix}, X = \begin{bmatrix} \underline{X} \\ \overline{X} \end{bmatrix}$$

In the Gauss Sidel method, we have:

$$\begin{bmatrix} D_1 + L_1 & 0 \\ S_2 & D_1 + L_1 \end{bmatrix}\begin{bmatrix} \underline{X} \\ \overline{X} \end{bmatrix} + \begin{bmatrix} U_1 & S_2 \\ 0 & U_1 \end{bmatrix}\begin{bmatrix} \underline{X} \\ \overline{X} \end{bmatrix} = \begin{bmatrix} \underline{Y} \\ \overline{Y} \end{bmatrix}$$

Then

$$\begin{aligned} \underline{X} &= (D_1 + L_1)^{-1}\underline{Y} - (D_1 + L_1)^{-1}U_1\underline{X} - (D_1 + L_1)^{-1}S_2\overline{X} \\ \overline{X} &= (D_1 + L_1)^{-1}\overline{Y} - (D_1 + L_1)^{-1}U_1\overline{X} - (D_1 + L_1)^{-1}S_2\underline{X} \end{aligned} \tag{2.14}$$

So, the Gauss Sidel iterative technique will be:

$$\begin{aligned} \underline{X}^{k+1} &= (D_1 + L_1)^{-1}\underline{Y} - (D_1 + L_1)^{-1}U_1\underline{X}^k - (D_1 + L_1)^{-1}S_2\overline{X}^k \\ \overline{X}^{k+1} &= (D_1 + L_1)^{-1}\overline{Y} - (D_1 + L_1)^{-1}U_1\overline{X}^k - (D_1 + L_1)^{-1}S_2\underline{X}^k \end{aligned} \tag{2.15}$$

The elements of $X^{k+1} = (\underline{X}^{k+1}, \overline{X}^{k+1})^t$ are

$$\underline{x}_i^{k+1} = \frac{1}{s_{i,j}} = \left[\underline{y}_i - \sum_{j=1}^{i-1} s_{i,j}\underline{x}_i^{k+1} - \sum_{j=i+1}^{n} s_{i,j}\underline{x}_i^{k} - \sum_{j=1}^{n} s_{i,n+j}\overline{x}_j^{k} \right]$$

$$\overline{x}_j^{k+1} = \frac{1}{s_{i,j}} = \left[\overline{y}_i - \sum_{j=1}^{i-1} s_{i,j}\overline{x}_j^{k+1} - \sum_{j=i+1}^{n} s_{i,j}\overline{x}_j^{k} - \sum_{j=1}^{n} s_{i,n+j}\underline{x}_i^{k} \right]$$

$k = 1, 2, \ldots i = 1, 2, \ldots, n$

The results in the matrix form of the Gauss Sidel iterative techniques are.
$X^{k+1} = PX^k + C$ where

$$P = \begin{bmatrix} -(D_1 + L_1)^{-1}U_1 & -(D_1 + L_1)^{-1}S_2 \\ -(D_1 + L_1)^{-1}S_2 & -(D_1 + L_1)^{-1}U_1 \end{bmatrix},$$

$$C = \begin{bmatrix} -(D_1 + L_1)^{-1}\underline{Y} \\ -(D_1 + L_1)^{-1}\overline{Y} \end{bmatrix}, X = \begin{bmatrix} \underline{X} \\ \overline{X} \end{bmatrix}$$

From Theorems (2.2.3) (2.2.2) both the Jacobi iterates and Gauss Sidel iterates are converge to the unique solution of $X = A^{-1}Y$, for any X^0, where $X \in R^{2n}$ and $(\underline{X}, \overline{X}) \in E^n$. The stopping criterion with tolerance $\epsilon > 0$ is

$$\frac{\overline{X}^{k+1} - \overline{X}^k}{\overline{X}^{k+1}} < \epsilon, \frac{\underline{X}^{k+1} - \underline{X}^k}{\underline{X}^{k+1}} < \epsilon, K = 0, 1, \ldots$$

2.2.4 Example

Consider the 2×2 interval system

$$\begin{cases} x_1 - x_2 = [0, 2], \\ x_1 + 3x_2 = [4, 7]. \end{cases}$$

The extended 4×4 matrix is

$$S = \begin{bmatrix} 1 & 0 & 0 & -1 \\ 1 & 3 & 0 & 0 \\ 0 & -1 & 1 & 0 \\ 0 & 0 & 1 & 3 \end{bmatrix}$$

$$X = \begin{bmatrix} \underline{x}_1 \\ \underline{x}_2 \\ \overline{x}_1 \\ \overline{x}_2 \end{bmatrix} = S^{-1}Y$$

$$= \begin{bmatrix} +0.1250 & -0.1250 & -0.3750 & +0.3750 \\ -0.3750 & -0.3750 & -0.3750 & -0.1250 \\ -0.3750 & +0.3750 & +0.3750 & -0.1250 \\ +0.1250 & -0.1250 & -0.3750 & +0.3750 \end{bmatrix}$$

The exact and solutions are

$$x_1 = \left[\underline{x}_1, \overline{x}_1\right] = [1.375, 2.875],$$
$$x_2 = \left[\underline{x}_2, \overline{x}_2\right] = [0.875, 1.375]$$

The Hausdorff distance of solutions with $\epsilon = 10^{-2}$, in the Jacobi method is 0.0027 and in the Gauss Sidel method is $2.31180e - 004$.

2.2.5 Example

Consider the 3×3 interval system

$$\begin{cases} 4x_1 + x_2 - x_3 = [0, 2] \\ -x_1 + 3x_2 + x_3 = [2, 3] \\ 2x_1 + x_2 + 3x_3 = [-2, -1] \end{cases}$$

The matrix of the system is strictly diagonally dominant. The extended 6×6 matrix is

$$S = \begin{bmatrix} 4 & 1 & 0 & 0 & 0 & -1 \\ 0 & 3 & 1 & -1 & 0 & 0 \\ 2 & 1 & 3 & 0 & 0 & 0 \\ 0 & 0 & -1 & 4 & 1 & 0 \\ -1 & 0 & 0 & 0 & 3 & 1 \\ 0 & 0 & 0 & 2 & 1 & 3 \end{bmatrix}$$

The elements of the solution vector are as follows:

$$x_1 = [-0.4125, 0.0351]$$
$$x_2 = [0.9125, 1.1076]$$
$$x_3 = [-0.6969, -0.7353]$$

By the fact that x_3 is not an interval number, The interval solution in this case is a weak solution given by

$$u_1 = [-0.4125, 0.0351]$$
$$u_2 = [0.9125, 1.1076]$$
$$u_3 = [-0.7353, -0.6969]$$

The Hausdorff distance of solution with $\epsilon = 10^{-3}$, in the Jacobi method is 0.0145 and in the Gauss Sidel method is 0.0139.

Chapter 3
Orthogonal Polynomials and Least Square Approximation

3.1 Orthogonal Polynomials

Different types of polynomials play important roles in applied and numerical mathematics. The Legendre, Laguerre and Hermit polynomials are introduced in the theory of integration just like the way that Chebyshev polynomials are used in the theory of approximation. All of these polynomials are orthogonal families and also satisfy certain second-order differential equations. Moreover, by using each three terms in successive order, a general recursive relationship can be defined for them. In the following, we will show how these properties are generally obtained when the interval as a domain and weight function are known.

3.1.1 Definition-Inner Product of Definite Functions

Suppose $[a, b]$ *is an arbitrary interval, and* $w(x)$ *is a definite positive function in this interval. The inner product of two definite functions* $p(x)$ *and* $q(x)$ *in the interval* $[a, b]$ *in continuous and discrete states, respectively, is defined by two following terms:*

$$p(x), q(x) = \int_a^b w(x)p(x) \cdot q(x)dx \tag{3.1}$$

$$p(x), q(x) = \sum_{n=1}^{N} w(x_n)p(x_n) \cdot q(x_n) \tag{3.2}$$

The second formula is out of the scope of this book. However, we assume in the first formula that for all the given functions of $p(x)$ and $q(x)$, this integral exists (at least one improper integral).

T. Allahviranloo and A. Esfandiari, *A Course on Integral Equations with Numerical Analysis*, Mathematical Engineering, https://doi.org/10.1007/978-3-030-85350-1_3

3.1.2 Definition-Orthogonal Functions

The set of functions (not necessarily polynomials) $\phi_0, \ldots, \phi_n$ is called orthogonal on the interval $[a, b]$ relative to the weight function $w(x)$ whenever:

$$\int_a^b w(x)\phi_j(x)\phi_k(x)dx = \begin{cases} 0, & j \neq k \\ \alpha_k > 0, & j = k \end{cases}$$

3.1.3 Example

Assuming $w(x) = 1$, it can be easily verified that functions $p(x) = 1$ and $q(x) = x$ are orthogonal on the interval $[-1, 1]$.

3.1.4 Example

Assuming $w(x) = \frac{1}{\sqrt{1-x^2}}$, functions $p(x) = 1$ and $q(x) = x$ are orthogonal on the interval $[-1, 1]$ because:

$$\begin{aligned} p(x), q(x) &= \int_{-1}^{1} \frac{x}{\sqrt{1-x^2}} dx \\ &= -\frac{1}{2} \cdot 2\left[\sqrt{1-x^2}\right]_{-1}^{1} = 0 \end{aligned}$$

3.1.5 Example

The functions $p(x) = \sin nx$ and $q(x) = \cos mx$, for the integers m and n, where $m \neq n$, are orthogonal on the interval $[0, 2\pi]$, relative to the weight function $\frac{1}{2\pi}$.

3.1.6 Definition

Sequence $\{p_i\}_{i=0}^{\infty}$ (not necessarily definite), where p_i's are real polynomials of exactly degree i, is called orthogonal when all p_i s are perpendicular to each other, in other words:

(1) *For each i*

$$p_i(x) = \alpha_i x^i + (polynomials\ with\ degree\ lower\ than\ i)$$

where, $\alpha_i \neq 0$.

(2) *For every* $i \neq j$, *we have*

$$p_i(x), p_j(x) = 0$$

3.1.7 Example

Suppose $p_0(x) = 1$, $p_1(x) = x$ and $p_2(x) = 3x^2 - 1$. Because

$$p_0(x), p_1(x) = 0$$

$$p_0(x), p_2(x) = \int_{-1}^{1} \left(3x^2 - 1\right) dx = 0$$

$$p_1(x), p_2(x) = \int_{-1}^{1} \left(3x^2 - 1\right) dx = 0$$

thus, p_0, p_1 and p_2 form an orthogonal sequence on the interval $[-1, 1]$.

Assume that all the following defined moments μ_k exist and are finite, and at least $\mu_0 > 0$, then

$$\mu_k = \int_a^b w(x) x^k dx, k = 0, 1, ... \quad (3.3)$$

In fact, we use these moments to determine the coefficients of a polynomial. Please note that all real polynomials form a vector space, and it can be easily shown that polynomials $1, x, x^2, x^3$ are linearly independent.

3.1.8 Theorem

If for every $j = 0, 1, ..., n$, $p_j(x)$ *is a polynomial of degree* j, *then* $\{p_0, ..., p_n\}$ *is linearly independent on each interval* $[a, b]$.

Proof Suppose that $\alpha_0, ..., \alpha_n$ are arbitrary real numbers such that for every x in the interval $[a, b]$ we have:

$$p(x) = \alpha_0 p_0(x) + \alpha_1 p_1(x) + \cdots + \alpha_n p_n(x) = 0$$

Then, we must prove that $\alpha_0 = \cdots = \alpha_n = 0$.

Since $p(x) = \sum_{j=0}^{n} \alpha_j p_j(x)$ becomes zero at any point on $[a, b]$, so the equation has $p(x) = 0$ infinite roots. On the other hand, we know that *p(x)* is a polynomial of degree n, so $p \equiv 0$ and the coefficient of each power of x is zero. Every p_j is of exactly degree j and therefore,

$$p(x) = \alpha_n x^n + \sum_{j=0}^{n-1} \alpha_j p_j(x)$$

requires that $\alpha_n = 0$. By continuing the same trend, it can be shown that:

$$\alpha_0 = \cdots = \alpha_n = 0$$

So $\{p_0, ..., p_n\}$ is linearly independent.

3.1.9 Orthogonal Polynomials and Least Squares Approximation

Suppose that $f \in c[a, b]$, and p_n is a polynomial of at most degree n that minimize

$$\|f - p_n\|_2 = \left(\int_a^b (f(x) - p_n(x))^2 dx \right)^{\frac{1}{2}}$$

To determine an approximating polynomial such as p_n, we suppose:

$$p_n(x) = a_n x^n + a_{n-1} x^{n-1} + \cdots + a_1 x + a_0 = \sum_{k=0}^{n} a_k x^k$$

and define:

$$E(a_0, ..., a_n) = \int_a^b \left(f(x) - \sum_{k=0}^{n} a_k x^k \right)^2 dx$$

The goal is to find the real coefficients $a_0, ..., a_n$ that minimize E. We know that the necessary and sufficient condition for E to be minimized by the numbers $a_0, ..., a_n$ is that:

$$\frac{\partial E}{\partial a_j} = 0, \ j = 0, ..., n$$

Because we can write:

$$E = \int_a^b (f(x))^2 dx - 2\sum_{k=0}^{n} a_k \int_a^b x^k f(x)dx + \int_a^b \left(\sum_{k=0}^{n} a_k x^k\right)^2 dx$$

then

$$\frac{\partial E}{\partial a_j} = -2\int_a^b x^j f(x)dx + 2\sum_{k=0}^{n} a_k \int_a^b x^{j+k} dx = 0$$

so to construct p_n, we must solve the following system of $n + 1$ linear equations with $n + 1$ unknowns:

$$\sum_{k=0}^{n} a_k \int_a^b x^{j+k} dx = \int_a^b x^j f(x)dx, \; j = 0, ..., n$$

The above equations are called normal equations. If $f \in c[a, b]$ and $a \neq b$, the system of equations always has a unique solution.

3.1.10 Example

To find the least squares approximating polynomials of degree 2 such as $p_2(x) = a_2x^2 + a_1x + a_0$, for the function $f(x) = \sin x$ on the interval $[a, b]$, the system of normal equations is as follows:

$$\begin{cases} a_0 \int_0^1 dx + a_1 \int_0^1 x dx + a_2 \int_0^1 x^2 dx = \int_0^1 \sin x dx \\ a_0 \int_0^1 dx + a_1 \int_0^1 x^2 dx + a_2 \int_0^1 x^3 dx = \int_0^1 x \sin x dx \\ a_0 \int_0^1 x^2 dx + a_1 \int_0^1 x^3 dx + a_2 \int_0^1 x^4 dx = \int_0^1 x^2 \sin x dx \end{cases}$$

After integration, we will have:

$$\begin{cases} a_0 + \frac{1}{2}a_1 + \frac{1}{3}a_2 = \frac{2}{\pi} \\ \frac{1}{2}a_0 + \frac{1}{3}a_1 + \frac{1}{4}a_2 = \frac{1}{\pi} \\ \frac{1}{3}a_0 + \frac{1}{4}a_1 + \frac{1}{5}a_2 = \frac{\pi^2-4}{\pi^3} \end{cases}$$

By solving the above-mentioned system, we have:

$$a_0 \simeq 0.050462, a_1 \simeq 4.12251, a_2 \simeq -4.12251$$

As a result:

$$p_2(x) = -4.12251x^2 + 4.12251x + 0.051462$$

But the above-mentioned method also has disadvantages, which we describe as follows:

(1) Solving $(n+1) \times (n+1)$ linear system is difficult for large n.
(2) Because the coefficients are in the form of

$$\int_a^b x^{j+k} dx = \frac{b^{j+k+1} - a^{j+k+1}}{j+k+1}$$

the resulted system does not have a proper numerical solution due to the rounding error, or better to say, the matrix can be ill-conditioned.
(3) We cannot generally claim that the calculating method that used for obtaining the best nth degree polynomial, i.e., p^n, is quite useful for calculating, too.

So, let us consider another method to approximate the least squares as follows. In this method, as soon as p_n is known, p_{n+1} is easily obtained.

Suppose Π_n be a set of all polynomials of at most degree n and Q_j represents polynomials of degree j. For set $\{Q_0, \ldots, Q_n\}$, the real numbers $a_0, \ldots, a_n$ must be calculated such that the following formula is minimized:

$$E(a_0, \ldots, a_n) = \int_a^b \left(f(x) - \sum_{k=0}^{n} a - kQ_k(x) \right)^2 dx$$

$$= \int_a^b (f(x))^2 dx - 2\sum_{k=0}^{n} a_k \int_a^b f(x)Q_k(x)dx$$

$$+ \sum_{j=0}^{n}\sum_{k=0}^{n} a_j a_k \int_a^b Q_j(x)Q_k(dx)dx$$

To do that, following condition is sufficient to be hold.

$$\frac{\partial E}{\partial a_j} = 0, \; j = 0, \ldots, n$$

By applying the above-mentioned condition, the normal equations will be shown as follows:

$$\sum_{j=0}^{n} a_j \int_a^b Q_j(x)Q_k(x)dx = \int_a^b f(x)Q_k(x)dx, \; k = 0, \ldots, n$$

Now if

$$\int_a^b Q_j(x)Q_k(x)dx = 0,\ j \neq k$$

Then, the normal equation system can be written as follows:

$$a_k \int_a^b (Q_k(x))^2 dx = \int_a^b f(x)Q_k(x)dx,\ k = 0, \ldots, n$$

Therefore,

$$a_k = \frac{\int_a^b f(x)Q_k(x)dx}{\int_a^b (Q_k(x))^2 dx},\ k = 0, \ldots, n$$

Given the Definition (3.1.6) and $w(x) = 1$ and the discussion above, we state the following theorem without proof.

3.1.11 Theorem

If $\{Q_0, \ldots, Q_n\}$ is an orthogonal set of functions on interval $[a, b]$ where $a < b$, then the least squares approximation of function f using $Q_0, \ldots, Q_n$ is:

$$p(x) = \sum_{k=0}^{n} a_k Q_k(x)$$

where

$$\begin{aligned} a_k &= \frac{\int_a^b f(x)Q_k(x)dx}{\int_a^b (Q_k(x))^2 dx} \\ &= \frac{1}{\alpha_k} \int_a^b Q_k f(x)dx \end{aligned}$$

3.1.12 Example

Suppose $f \in c[-\pi, \pi]$, and the set of functions $\{Q_0, \ldots, Q_n\}$ where

$$Q_0(x) = \frac{1}{\sqrt{2\pi}}$$
$$Q_{2k}(x) = \frac{1}{\sqrt{\pi}} \cos kx, k = 1, ..., n$$
$$Q_{2k-1}(x) = \frac{1}{\sqrt{\pi}} \sin kx, k = 1, ..., n$$

form an orthogonal set of functions on interval $[-\pi, \pi]$. It is obvious that:

$$\int_{-\pi}^{\pi} (Q_k(x))^2 dx = 1, k = 0, 1, ..., 2n$$

The least squares approximation, which is a trigonometric polynomial, can be defined as

$$T_n(x) = \sum_{k=0}^{n} a_k Q_k(x)$$

where

$$a_k = \int_{-\pi}^{\pi} f(x) Q_k(x) dx, k = 0, ..., 2n$$

Now if $f(x) = |x|$, the coefficients of above-mentioned formula are calculated as follows:

$$a_0 = \int_{-\pi}^{\pi} |x| \frac{1}{\sqrt{2\pi}} dx = \frac{\pi^2}{\sqrt{2\pi}}$$
$$a_{2k}(x) = \int_{-\pi}^{\pi} |x| \cos kx dx = \frac{2}{k^2\sqrt{\pi}}((-1)^k - 1)$$
$$a_{2k-1}(x) = \frac{1}{\sqrt{\pi}} \int_{-\pi}^{\pi} |x| \sin kx dx = 0$$

Therefore, the trigonometric polynomial that approximates f is in the following form:

$$T_n(x) = \frac{\pi}{2} - \frac{2}{\pi} \sum_{k=0}^{n} \frac{(-1)^k - 1}{k^2} \cos kx$$

Chapter 4
Integral Equations

4.1 Introduction

Generally, the integral equation is an equation in which the unknown function is appeared under the integral signs. Based on many applications of these equations, it is needed to discuss several methods to solve them. Sometimes the unknown function is found analytically and numerically. In this chapter, we will discuss these methods.

4.1.1 Definition—Integral Equation

Whenever in an equation, the unknown function is appeared in both sides of the equation and specially, inside of the integral sign, it is called an integral equation.

4.1.2 Example

(1) $y(s) = \int_a^b k(s,t)x(t)dt, \quad a \le s \le b$

(2) $x(s) = y(s) + \int_a^b k(s,t)x(t)dt, \quad a \le s \le b$

(3) $x(s) = \int_a^b k(s,t)x^2(t)dt, \quad a \le s \le b$

(4) $x(s) = y(s) + \int_a^b k(s,t,x(t))dt, \quad a \le s \le b$

(5) $x(s) = y(s) + \int_a^b \int_a^b k(s,t)x(t)x(u)dtdu, \quad a \le s \le b.$

T. Allahviranloo and A. Esfandiari, *A Course on Integral Equations with Numerical Analysis*, Mathematical Engineering, https://doi.org/10.1007/978-3-030-85350-1_4

In each equation of Example (4.1.2), function $x(\cdot)$ is unknown. Equations (1) and (2) can also be written in functional form:

$$L(x(s)) = y(s) \tag{4.1}$$

where L is a linear operator in the following form that comprises the unknown function; i.e., it is the operator that acts on the unknown function. For example, for (1), L is the same integration, and for (2), L is defined as

$$L : I - \int_a^b$$

Sometimes the operator L can also be represented as $L := I - K$, where I is the identity matrix and k is arbitrary matrix such as function $k(s, t)$. Therefore, the operator L is also called matrix operator or matrix representation. Obviously, L has linear properties. In other words, it holds the linearity of each linear combination. That means that we have:

$$\forall \lambda_i \left(\lambda_i \in \mathbb{R} \;\&\; i = 1, \ \ldots \ n \Rightarrow L\left(\sum_{i=1}^{n} \lambda_i x_i(s) \right) = \sum_{i=1}^{n} \lambda_i L(x_i(s)) \right)$$

So using this notation, for Eq. (1), we have:

$$L(x(s)) = \int_a^b k(s, t)x(t)dt$$

And for Eq. (2), we have:

$$L(x(s)) = x(s) - \int_a^b k(s, t)x(t)dt$$

The equations in the form of (4.1.2), where $L(x(s))$ is a linear expression in terms of $x(s)$, and $x(s)$ is under one or more integral signs, are called Fredholm linear integral equations. Therefore, Eqs. (1) and (2) are linear and Eqs. (3), (4) and (5) are nonlinear. In this section, we will only study linear equations.

4.1.3 Definition—First Type Integral Equation

The integral equations, such as Eq. (1), where $x(s)$ (unknown function) appears only under the integral sign and does not appear anywhere else, are called integral equations of the first type.

Also, equations such as Eq. (2), where the unknown function is both inside and outside the integral, are called integral equations of the second type.

4.1.4 Definition—Kernel

In each equation of (1) and (2), the function $k(s, t)$ is called the kernel of the equation, which is a square plane for every $a \leq s \leq b$ and every $a \leq t \leq b$.

Obviously, in each of equations that are presented in Example (4.1.2), the kernel function $k(s, t)$ and $y(s)$ are known.

4.1.5 Definition—Homogeneous Integral Equation of the Second Type

If in the second type of equation, like Eq. (2), we have:

$$L(x(s))$$

That is, $y(s) = 0$, in which case, the resulting equation is a homogeneous integral equation of the second type, i.e.,

$$x(s) = \int_a^b k(s, t)x(t)dt$$

If in each equation of Example (4.1.2), we substitute the variable s in the upper bound of the integral, instead of b, then the **Volterra integral equations** are produced. Therefore, the equations are rewritten as follows:

(6) $y(s) = \int_a^s k(s, t)x(t)dt, \quad a \leq s \leq b$

(7) $x(s) = y(s) + \int_a^s k(s, t)x(t)dt, \quad a \leq s \leq b$

(8) $x(s) = \int_a^s k(s, t)x(t)dt, \quad a \leq s \leq b$

(9) $x(s) = \int_a^s k(s, t)x^2(t)dt, \quad a \leq s \leq b$

(10) $x(s) = y(s) + \int_a^s k(s, t, x(t))dt, \quad a \leq s \leq b.$

As before, the linear, nonlinear, homogeneous, first-type and second-type categories are also hold for (6)–(10). Equation (6) is linear Volterra equation of first type, Eq. (9) is nonlinear Volterra equation of first type, Eqs. (7) and (10) are linear

Volterra equations of second type and Eq. (8) is homogeneous linear Volterra equation of second type.

The mentioned contents can be expressed another way in the form of the following definition:

4.1.6 Definition—Volterra Integral Equation

In the Fredholm integral equation, if we consider:

$$k(s,\ t) = 0,\ t > s$$

then the Volterra integral equation is obtained.

$$x(s) = f(s) + \lambda \int_a^b k(s,t)x(t)dt$$

4.1.7 Definition—Integro-Differential Equation

An equation is called an integro-differential equation, if

(A) The unknown function is under the integral sign.
(B) The unknown function is under the differential sign.
(C) There is an initial condition for the unknown function.

4.1.8 Definition—The Integro-Differential Equation

Equation

$$\begin{cases} x'(t) + p(x)x(t) + y(t) = \int_a^b k(s,u)x(u)du \\ x(a) = y_0 \end{cases} \tag{4.2}$$

is called the integro-differential equation.

In the following, we will explain the numerical methods for solving such equations in detail.

4.1.9 The Relationship Between Integral Equations and Differential Equations

In this section, we want to express the relationship between the integral equation and the differential equation. Also, we will see that the solution of a differential equation with an initial value is a solution of Volterra integral equation. The differential equation can be assumed to be the first order with an initial condition or the second order with two initial conditions; in both cases, the answer will be a Volterra integral equation with one integral sign and two integral signs, respectively. On the other hand, the equations in which there are both differential and integral operators are also examined.

Before expressing this relationship, we express and prove the following lemma.

4.1.10 Lemma

Suppose F is a continuous function, then, if

$$y(x) := y = \int_0^x F(x,\ u)du$$

Then

$$y'_x = F(x,\ x) + \int_0^x F'_x(x,\ u)du$$

Proof First, we form $y(x + \Delta x) - y(x)$:

$$\begin{aligned} y(x+\Delta x) - y(x) &= \int_0^{x+\Delta x} F(x+\Delta x,\ u)du - \int_0^x F(x,\ u)du \\ &= \int_0^x F(x+\Delta x,\ u)du + \int_x^{x+\Delta x} F(x+\Delta x,\ u)du \\ &\quad - \int_0^x F(x,\ u)du \\ &= \int_0^x (F(x+\Delta x,\ u) - F(x,\ u))du \\ &\quad + \int_x^{x+\Delta x} F(x+\Delta x,\ u)du \end{aligned}$$

Now, we multiply both sides of the above relation by $\frac{1}{\Delta x}$:

$$\frac{y(x+\Delta x)-y(x)}{\Delta x}=\int_0^x \frac{F(x+\Delta x,u)-F(x,u)}{\Delta x}du + \int_x^{x+\Delta x} \frac{F(x+\Delta x,u)}{\Delta x}du$$

Given that $F(x+\Delta x, u)$ is a continuous function on the integration interval, and also because the constant function does not change the sign on this interval, so according to the integral mean value theorem, we have:

$$\exists\eta(\eta \in [x,\ x+\Delta x]\ \&\ \frac{1}{\Delta x}\int_x^{x+\Delta x} F(x+\Delta x,\ u)du = F(x+\Delta x,\ \eta)\frac{1}{\Delta x}\int_x^{x+\Delta x} 1\,du$$

Therefore, we will have:

$$\frac{y(x+\Delta x)-y(x)}{\Delta x}=\int_0^x \frac{F(x+\Delta x,u)-F(x,u)}{\Delta x}du + F(x+\Delta x,\eta)$$

If we take the limit of the both sides of the above relationship, we will have:

$$\lim_{\Delta x\to 0}\frac{y(x+\Delta x)-y(x)}{\Delta x}=\int_0^x \lim_{\Delta x\to 0}\frac{F(x+\Delta x,u)-F(x,u)}{\Delta x}du + \lim_{\Delta x\to 0} F(x+\Delta x,\eta)$$

So, it can be said that:

$$y'_x=\int_0^x F'(x,\ u)du + F(x,\ x)$$

Now, according to the above-mentioned fundamentals, we examine the relationship between the first-order differential equation and the integral equation. Suppose that

$$\begin{cases} y'(x) = f(x,\ y(x)) \\ y(0) = y_0 \end{cases} \tag{4.3}$$

where $f(x, y)$ is continuous with respect to (x, y). By integrating both sides of the relation (4.3) on the interval $[0,\ x]$ and using the change of variables, we have:

$$\int_0^x y'(t)dt = \int_0^x f(t,\ y(t))dt$$

In this case

$$y(x) = y_0 + \int_0^x f(t,\ y(t))dt \tag{4.4}$$

In fact, we proved the following lemma.

4.1.11 Lemma

Suppose that $f(x,\ y)$ is continuous with respect to $(x,\ y)$. In this case, the solution of Eq. (4.3) is a nonlinear Volterra integral equation of the second type.

Now, we consider the second-order differential equation and examine its relationship with the integral equation. Suppose that

$$\begin{cases} \frac{d^2y}{dx^2} = y''(x) = f(x,\ y(x)) \\ y(0) = y_0,\ y'(0) = y_1 \end{cases} \tag{4.5}$$

As before, if we integrate both sides of Eq. (4.5) on the interval $[0, x]$, we will have:

$$y'(x) = y_1 + \int_0^x f(t,\ y(t))dt$$

By substituting u instead of x, we have:

$$y'(u) = y_1 + \int_0^u f(t, y(t))dt. \tag{4.6}$$

Multiplying Eq. (4.6) by du and reintegrating on the interval $[0, x]$, we have:

$$\int_0^x y'(u)du = \int_0^x y_1 du + \int_0^x \int_0^u f(t,\ y(t))dtdu$$

Therefore, the following relation is obtained:

$$y(x) = y_0 + y_1 x + \int_0^x \int_0^u f(t,\ y(t))dtdu. \tag{4.7}$$

Given that $0 \le t \le u \le x$, we claim:

$$\begin{aligned} g(x) &= \int_0^x \int_0^u f(t,\ y(t))dtdu \\ &= \int_0^x f(t,\ y(t))dt \int_t^x du \end{aligned}$$

$$= \int_0^x (x - t) f(t, \ y(t)) dt$$
$$= h(x)$$

To prove this claim ($g(x) = h(x)$), it is sufficient to prove $g'(x) = h'(x)$, because in this case

$$g(x) = h(x) + c$$

and since $h(0) = g(0)$, so $c = 0$.

According to Lemma (4.1.11), we have:

$$g'(x) = \int_0^x f(t, \ y(t)) dt$$

Because, by assuming

$$F(y, \ u) = \int_0^u f(t, \ y(t)) dt$$

we have:

$$\begin{aligned} g'(x) &= F(y, \ x) + \int_0^x F'_y(y, \ u) du \\ &= \int_0^x f(t, \ y(t)) dt + 0 \end{aligned} \tag{4.8}$$

Therefore, according to Lemma (4.1.11), we have:

$$h'(x) = \int_0^x f(t, \ y(t)) dt$$

Because, assuming

$$F(x, \ t) = (x - t) f(t, \ y(t))$$

We have

$$\begin{aligned} h'(x) &= F(x, \ x) + \int_0^x F'(x, \ t) dt \\ &= 0 + \int_0^x f(t, \ y(t)) dt \end{aligned} \tag{4.9}$$

So, according to Eqs. (4.8) and (4.9), the sentence is true. Therefore, Eq. (4.7) is obtained as follows:

$$y(x) = y_0 + y_1 x + \int_0^x (x - t) f(t, y(t)) dt \tag{4.10}$$

which is a nonlinear Volterra integral equation of the second type.

The above discussion can be stated by the following lemma.

4.1.12 Lemma

The solution of the second-order differential equation with the boundary conditions (4.5) is a nonlinear integral equation of the second type.

We now consider the following second-order differential equation:

$$\begin{cases} y''(x) = f(x,\ y(x)) \\ y(0) = \alpha,\ y(l) = \beta \end{cases} \quad 0 \le x \le l \tag{4.11}$$

First, according to Lemma (4.1.12), the general solution for Eq. (4.11) is as follows

$$y(x) = A + Bx + \int_0^x (x - t) f(t,\ y(t)) dt \tag{4.12}$$

Obviously, Eq. (4.12) holds both in the differential equation and in the boundary conditions, i.e.,

$$y(0) = \alpha \Rightarrow \alpha = A$$

Therefore, A is obtained and

$$y(l) = \beta = \alpha + Bl + \int_0^l (l - t) f(t,\ y(t)) dt \tag{4.13}$$

So, by obtaining B from Eq. (4.13) and substituting it in Eq. (4.12), we have:

$$\begin{aligned} y(x) &= \alpha + \frac{(\beta - \alpha)x}{l} + \int_0^x (x - t) f(t,\ y(t)) dt \\ &\quad - \frac{x}{l} \int_0^l (l - t) f(t,\ y(t)) dt \\ &= \alpha + \frac{(\beta - \alpha)x}{l} + \int_0^x (x - t) f(t,\ y(t)) dt \end{aligned}$$

$$-\frac{x}{l}\int_0^x (l-t)f(t,\ y(t))dt - \frac{x}{l}\int_x^l (l-t)f(t,\ y(t))dt$$

$$= \alpha + \frac{(\beta-\alpha)x}{l} + \int_0^x \left((x-t) - \frac{x(l-t)}{l}\right) f(t,\ y(t))dt$$

$$-\int_x^l \frac{x}{l}(l-t)f(t,\ y(t))dt$$

$$= \alpha + \frac{(\beta-\alpha)x}{l} + \int_0^x \frac{t(x-l)}{l} f(t,\ y(t))dt$$

$$+\int_x^l \frac{x}{l}(t-l)f(t,\ y(t))dt$$

The above equation can be written as follows:

$$y(x) = z(x) + \int_0^l k(x,\ t)f(t,\ y(t))dt \tag{4.14}$$

where

$$k(x,\ t) = \begin{cases} \frac{t(x-l)}{l}, & 0 \le t \le x \\ \frac{x(t-l)}{l}, & x \le t \le l \end{cases}$$

and

$$z(x) = \alpha + \frac{\beta - \frac{\alpha}{x}}{l}, \quad \alpha,\ \beta \in \mathbb{R}$$

Obviously, (4.14) is also a nonlinear Fredholm integral equation of second type. As it can be seen, the integral kernel, $k(x, t)$, holds in the following equation

$$k(x,\ t) = k(t,\ x) \tag{4.15}$$

4.1.13 Definition

If Eq. (4.15) holds in an integral equation, the kernel is called symmetric.

Now the question is that in such cases, how can an asymmetric kernel be transformed into a symmetric kernel?

Suppose

$$y''(x) + p(x)y(x) = w(x), \quad p(x) \ge 0$$

in this case

$$\begin{cases} y''(x) = w(x) - p(x)y(x) = f(x, y(x)) \\ y(0),\ y(l) \text{ are known} \end{cases}$$

According to Eq. (4.14), we have:

$$\begin{aligned} y(x) &= z(x) + \int_0^l k(x,\ t)(w(t) - p(t)y(t))dt \\ &= z(x) + \int_0^l k(x,\ t)w(t)dt - \int_0^l k(x,\ t)p(t)y(t)dt \end{aligned}$$

So, assuming

$$T(x) = z(x) + \int_0^l k(x,\ t)w(t)dt$$

we have:

$$y(x) = T(x) - \int_0^l k(x,\ t)p(t)y(t)dt$$

where $k(x,\ t)p(t)$ is an asymmetric kernel but $k(x,\ t)$ is the same symmetric kernel. If we multiply both sides of the above equation by $\sqrt{p(x)}$, we will have:

$$\sqrt{p(x)}y(x) = \sqrt{p(x)}T(x) - \int_0^l \sqrt{p(x)}k(x,\ t)\sqrt{p(x)}\sqrt{p(x)}y(t)dt$$

Now, assuming

$$\sqrt{p(t)}y(t) = \underline{y}(t)$$

we have:

$$\underline{y}(x) = \sqrt{p(x)}y(x) = \sqrt{p(x)}T(x) - \int_0^l \sqrt{p(x)}k(x,\ t)\sqrt{p(x)}\underline{y}(x)dt$$

Obviously, the above equation is a Fredholm integral equation of the second type that has a symmetric kernel $\sqrt{p(x)}k(x,\ t)\sqrt{p(t)}$.

Above discussions about how to convert kernels to the symmetric kernels hold exclusively for the differential Eq. (4.14), and it cannot be generalized to integral equations, in general.

4.2 Continuous Functions $x(.)$ and L^2

Since a necessary condition for integration is the continuity of the function under the integral sign, so we must discuss the continuity of each of functions $x(t)$ and $k(s, t)$. Also, for the integral of the integral equation to exist, the above functions must be located in space L^2. In this section, we assume that $x(t)$ is a complex function in terms of t on the interval $[a, b]$.

4.2.1 Definition

The L^2-space comprises of continuous and measurable functions whose square integrals are available, namely

$$L^2(a, b) = \left\{ x(t) | \int_a^b |x(t)|^2 dt < \infty \right\}$$

4.2.2 Definition

The two functions $x(t)$ and $y(t)$ of L^2-space are almost equal everywhere if the size of the set $A = \{t|x(t) \neq y(t)\}$ is equal to zero, i.e., $m(A) = 0$, and we can write $x(t) =^0 y(t)$ and if $m(B) = 0$ where

$$B = \{t|x(t) \neq 0\}$$

In this case, the function $x(t)$ is almost equal to zero everywhere, and we can write $x(t) =^0 0$. We also can say that $x(t)$ is not more than $y(t)$ everywhere, if the size of the set

$$C = \{t|x(t) \nleq y(t)\}$$

is equal to zero, i.e., $m(C) = 0$ and we can write $x(t) \leq^0 y(t)$.

4.2.3 Definition

If $x \in L^2$, then $\|x\|_2$ can be defined as follows:

$$\|x\|_2 = \left(\int_a^b |x(t)|^2 dt\right)^{\frac{1}{2}}$$

If $\|x\|_2 = 0$, then $x(t) =^0 0$ and vice versa.
So, we can say that the above-mentioned norm is a quasi-norm.

4.2.4 *Definition*

If $x(t)$ is a continuous function on the interval $[a, \ b]$, then we define $\|x\|_c$ as follows:

$$\|x\|_c = \sup_{a \le t \le b} |x(t)|$$

If $\|x\|_c = 0$, then $x(t) = 0$ and vice versa.
The above-mentioned norm is also called continuous norm of the function x.
Now, according to the above discussion, we define the Cauchy–Schwarz inequality theorem.

4.2.5 *Cauchy–Schwarz Inequality Theorem*

If $x(t)$ and $y(t)$ are functions in the L^2-space, then $x(t) \cdot y(t)$ is integrable and

$$\left|\int_a^b x(t)y(t)dt\right| \le \|x\| \cdot \|y\| \tag{4.16}$$

Proof Obviously, the function $x(t) \cdot y(t)$ is measurable. Given that

$$\left|\int_a^b x(t) \ \cdot \ y(t)dt\right| \le \int_a^b |x(t)| \ \cdot \ |y(t)|dt$$

and since the right integral of the above inequality exists, so the left integral also exists. To prove it, it suffices to consider $x(t)$ and $y(t)$ as real and negative functions, in which case we have:

$$x(t) \ \cdot \ y(t) \le \frac{1}{2}\left(x^2(t) + y^2(t)\right)$$

It is easy to obtain the integrability of $x(t) \ \cdot \ y(t)$ from the above equation. For every real λ, we have the following equation:

$$0 \leq \int_a^b (\lambda x(t) + y(t))^2 dt$$
$$= \lambda^2 \int_a^b x^2(t)dt + 2\lambda \int_a^b x(t) \cdot y(t)dt + \int_a^b y^2(t)dt$$
$$= \lambda^2 \|x\|^2 + 2\lambda \int_a^b x(t) \cdot y(t)dt + \|y\|^2$$

The above equation is a quadratic equation in terms of λ. In order for the above inequality to exist, Δ or Δ' must be negative, because $\|x\|^2 > 0$, therefore

$$\Delta' = \left(\int_a^b x(t) \cdot y(t)dt\right)^2 - \|x\|^2 \cdot \|y\|^2 < 0$$

and as a result:

$$\left|\int_a^b x(t)y(t)dt\right| \leq \|x\| \cdot \|y\|$$

The left side of the equation, i.e., $\int_a^b x(t)y(t)dt$, shows the same inner product of the functions $x(t)$ and $y(t)$, which in the more general case can be written as follows

$$\langle x, y\rangle = \int_a^b x(t) \cdot \overline{y(t)}dt$$

where $\overline{y(t)}$ is the complex conjugate of the function $y(t)$. As a result:

$$\overline{|\langle x, y\rangle|} = \leq \|x\| \cdot \|y\| \tag{4.17}$$

Some properties of inner product can also be easily proved. For example,

$$(I)\ \overline{|\langle x, y\rangle|} = \overline{\int_a^b x(t) \cdot \overline{y(t)}dt}$$
$$= \int_a^b \overline{x(t)} \cdot y(t)dt$$
$$= \langle y, x\rangle$$

and

$$(II)\ \langle x, y\rangle = \int_a^b x(t) \cdot \overline{x(t)}dt$$

$$= \int_a^b |x(t)|^2 dt$$
$$= \|x\|^2$$

Equation (4.17) can also be easily proven, too.

$$|\langle x,\ y\rangle| = \left|\int_a^b x(t) \cdot \overline{y(t)} dt\right|$$
$$\leq \int_a^b |x(t)| \cdot \left|\overline{y(t)}\right| dt$$
$$= \int_a^b |x(t)| \cdot |y(t)| dt$$
$$\leq \|x\| \cdot \|y\|$$

also

$$(III)\ \left\langle x, \sum_{i=1}^{n} \lambda_i y_i \right\rangle = \int_a^b x(t) \overline{\sum_{i=1}^{n} \lambda_i y_i(t) dt}$$
$$= \sum_{i=1}^{n} \overline{\lambda}_i \int_a^b x(t) \overline{y_i(t)} dt$$
$$= \sum_{i=1}^{n} \overline{\lambda}_i x,\ y_i$$

Obviously, according to *(I)*, if we have $\langle x,\ y\rangle = 0$, then $\langle y,\ x\rangle = 0$. If the inner product of two functions is zero, we say that these two functions are perpendicular to each other. Therefore, it can be said that the orthogonal property is a symmetric property.

4.2.6 Theorem

If $x(t)$ and $y(t)$ are functions in the L^2-space, then $x(t) + y(t)$ is also a function in the space L^2 and we have:

$$\|x + y\| \leq \|x\| + \|y\| \tag{4.18}$$

Proof It is enough to prove:

$$\|x + y\|^2 \leq (\|x\| + \|y\|)^2$$

Therefore, we have:

$$\begin{aligned}
\|x+y\|^2 &= \int_a^b (x+y)\overline{(x+y)}dt \\
&= \int_a^b |x(t)+y(t)|^2 dt \\
&\le \int_a^b (|x(t)|+|y(t)|)^2 dt \\
&= \int_a^b |x(t)|^2 dt + \int_a^b |y(t)|^2 dt + 2\int_a^b |x(t)| \cdot |y(t)| dt \\
&\le \|x\|^2 + \|y\|^2 + 2\|x\| \cdot \|y\|
\end{aligned}$$

so, we have:

$$\|x\| + \|y\|^2 \le (\|x\| + \|y\|)^2$$

Using Definition (4.2.4), we can easily prove the Theorem (4.2.6) for $\|\cdot\|_c$ as well, i.e.,

$$\|x+y\|_c \le \|x\|_c + \|y\|_c \tag{4.19}$$

The relationship between inner product and continuous norm can also be expressed and proved as follows. We have:

$$\begin{aligned}
\left|\int_a^b x(t) \cdot y(t) dt\right| &\le \int_a^b |x(t)| \cdot |y(t)| dt \\
&\le \int_a^b \|x\|_c \cdot \|y\|_c dt \\
&= (b-a)\|x\|_c \cdot \|y\|_c
\end{aligned}$$

therefore

$$|\langle x, y\rangle| \le (b-a)\|x\|_c \cdot \|y\|_c$$

According to Eq. (4.19) and Theorem (4.2.6), it can be claimed that if $x(t)$ and $y(t)$ are functions of L^2, then each linear combination of them, such as

$$\lambda x(t) + \mu y(t), \quad \lambda,\ \mu \in \mathbb{R}$$

is also a function of L^2. So, all functions of L^2 form a complex vector space. This is also true for continuous functions. Another property that can be mentioned for the inner product of two functions is that the inner product of the two functions is linear in terms of first components but is nonlinear in terms of second components, i.e.,

$$\left\langle \sum_{i=1}^{n} \lambda_i x_i,\ y \right\rangle = \sum_{i=1}^{n} \lambda_i \langle x_i,\ y \rangle$$

$$\left\langle x, \sum_{i=1}^{n} \lambda_i x_i \right\rangle = \sum_{i=1}^{n} \overline{\lambda_i} \langle x,\ y_i \rangle$$

Let us now check the continuity of the kernels.

Suppose that $k(s, t)$ is a continuous function of (s, t) on a square of $a \le s \le b$ and $a \le t \le b$. If $x(t)$ is a continuous function of t, then

$$y(s) = \int_a^b k(s, t)x(t)dt, \quad a \le s \le b \tag{4.20}$$

is also a continuous function, i.e.,

$$x \in C[a, b] \;\&\; k \in C([a, b] \times [a, b])$$

in this case

$$y = kx \in C[a, b]$$

where $y = kx$ is the compact form of Eq. (4.20). From the continuity of $x(t)$, we have:

$$\forall \varepsilon \exists \delta \forall t \forall t' \left(|t - t'| < \delta \Rightarrow |x(t) - x(t')| < \varepsilon \right)$$

Also, from the continuity of $k(s,\ t)$, we have:

$$\forall \varepsilon \exists \delta \forall (s,\ t) \forall (s',\ t') \left(d((s,\ t), (s',\ t')) < \delta \Rightarrow |k(s,\ t) - k(s',\ t')| < \varepsilon \right)$$

It is enough to show that

$$\forall \varepsilon \exists \delta \forall s \forall s' \left(|s - s'| < \delta \Rightarrow |y(s) - y(s')| < \varepsilon \right)$$

For this purpose, we write:

$$\begin{aligned} |y(s) - y(s')| &= \left| \int_a^b k(s,\ t)x(t)dt - \int_a^b k(s',\ t)x(t)dt \right| \\ &= \left| \int_a^b (k(s, t) - k(s', t))x(t)dt \right| \end{aligned}$$

Given the continuity of thee function $x(t)$ and the compactness of $[a,\ b]$, it can be said that this function is bounded. That means:

$$\exists m \forall t (|x(t)| \leq m)$$

Therefore, according to the continuity of $k(s, t)$, we can write:

$$d\big((s, t), (s', t')\big) = \sqrt{(s - s')^2 + (t - t')^2} = |s - s'| < \delta$$

and

$$|k(s, t) - k(s', t)| < \varepsilon',\ \varepsilon' = \frac{\varepsilon}{(b - a)m}$$

Therefore,

$$\begin{aligned}
|y(s) - y(s')| &\leq \left| \int_a^b \big(k(s, t) - k(s', t)\big) x(t) dt \right| \\
&\leq \int_a^b \varepsilon' \cdot m dt \\
&= \varepsilon' \cdot m \cdot (b - a) = \varepsilon
\end{aligned}$$

4.2.7 *Definition—Continuous Norm of a Continuous Kernel*

We define a continuous norm of a continuous kernel $k(s, t)$ as follows.

$$\|k\|_c = (b - a) \sup_{a \leq s,\ t \leq b} |k(s, t)|$$

It can be shown that the continuous kernel satisfies the following property.

$$\|y\|_c \leq \|k\|_c \cdot \|x\|_c$$

because

$$y(s) = \int_a^b k(s, t) x(t) dt$$

therefore

$$\begin{aligned}|y(s)| &\le \int_a^b |k(s,\ t)| \cdot |x(t)|dt \\ &\le \int_a^b \sup_{s,\,t}|k(s,\ t)| \cdot \|x\|_c dt \\ &= (b-a)\sup_{s,\,t}|k(s,\ t)| \cdot \|x\|_c \\ &= \|k\|_c \cdot \|x\|_c\end{aligned}$$

Now, we have

$$\forall s(|y(s)| \le \|k\|_c \cdot \|x\|_c)$$

that means

$$\sup_{s,\,t}|y(s)| \le \|k\|_c \cdot \|x\|_c$$

thus

$$\|y\|_c \le \|k\|_c \cdot \|x\|_c$$

In fact, we proved that:

$$y = kx \Rightarrow \|y\|_c \le \|k\|_c \cdot \|x\|_c$$

4.2.8 Definition—Linear Operator

The operator k is called linear, if we have:

$$\forall\lambda_i\left(k\left(\sum_{i=1}^{n}\lambda_i x_i\right) = \sum_{i=1}^{n}\lambda_i k x_i\right),\quad 1 \le i \le n \tag{4.21}$$

Equation (4.21) can be easily proved.

$$\begin{aligned}k\left(\sum_{i=1}^{n}\lambda_i x_i\right) &= \int_a^b k(s,\ t)\left(\sum_{i=1}^{n}\lambda_i x_i(t)\right)dt \\ &= \sum_{i=1}^{n}\lambda_i k x_i\end{aligned}$$

4.3 Production of Two Kernels

Suppose that $y = kx$, i.e.,

$$y(s) = \int_a^b k(s, t)x(t)dt, \quad a \le s \le b$$

If h is another operator such that $I = hy$, and

$$I(r) = \int_a^b h(r, s)y(s)ds$$

in this case

$$\begin{aligned} I(r) &= \int_a^b h(r, s)(k(s, t)x(t)dt)ds \\ &= \int_a^b \int_a^b h(r, s)k(s, t)x(t)dtds \\ &= \int_a^b \left(\int_a^b h(r, s)k(s, t)ds \right) x(t)dt \end{aligned}$$

we assume that

$$I(r) = \int_a^b l(r, t)x(t)dt$$

where

$$\begin{aligned} l(r, t) &= \int_a^b h(r, s)k(s, t)ds \\ &= (hk)(r, t) \end{aligned} \tag{4.22}$$

Equation (4.22) defines the combination or product of two kernels. Then

$$l = hk$$

4.3.1 Lemma

Suppose that k and h are two continuous functions on a square of $a \leq t$ and $s \leq b$, in this case

$$\|hk\|_c \leq \|h\|_c \cdot \|k\|_c \tag{4.23}$$

Proof According to Eq. (4.22), we have:

$$\begin{aligned}
|l(r,\ t)| &\leq \int_a^b |h(r,\ s)| \cdot |k(s,\ t)| ds \\
&\leq \int_a^b \sup_{r,\,s} |h(r,\ s)| \cdot \sup_{s,\,t} |k(s,\ t)| ds \\
&= (b-a) \sup_{r,\,s} |h(r,\ s)| \cdot \sup_{s,\,t} |k(s,\ t)| \\
&= \|h\|_c \cdot \frac{b-a}{b-a} \cdot \sup_{s,\,t} |k(s,\ t)| \\
&= \frac{1}{b-a} \|h\|_c \cdot \|k\|_c
\end{aligned}$$

Then

$$(b-a)|l(r,\ t)| \leq \|h\|_c \cdot \|k\|_c$$

Given that the above equation holds for every r and every t, therefore

$$(b-a) \sup_{r,\,t} |l(r,\ t)| \leq \|h\|_c \cdot \|k\|_c$$

as a result

$$\|hk\|_c \leq \|h\|_c \cdot \|k\|_c$$

It can be easily shown that the combination or product of kernels is not commutative. That means:

$$hk \neq kh$$

Because if we suppose $[a,\ b] = [0,\ 1]$, $k(s,\ t) = s^2 t$ and $h(r,\ s) = rs^2$, then:

$$(hk)(r,\ t) = \int_0^1 h(r,\ s) k(s,\ t) ds$$

$$= \int_0^1 rs^2 \cdot s^2 t ds$$
$$= \frac{1}{5} r \cdot t$$

But

$$(hk)(r,\ t) = \int_0^1 k(r,\ s)h(s,\ t)ds$$
$$= \int_0^1 r^2 s \cdot st^2 ds$$
$$= \frac{1}{3} r^2 \cdot t^2$$

So, in general, we can say that $hk \neq kh$, unless for $t = 0$. It can also be shown that

(1) $(hk)l = h(kl)$
(2) $h(k + l) = hk + hl$
(3) $(k + l)h = kh + lh$

We prove the part (1) and leave the proof of the other two parts to the reader. We know:

$$[(hk)l](r, t) = \int_a^b (hk)(r, s)l(s, t)ds$$

On the other hand:

$$(hk)(r, s) = \int_a^b h(r, z)k(z, s)dz$$

So, we have:

$$[(hk)l](r,\ t) = \int_a^b \left(\int_a^b h(r,\ z)k(z,\ s)dz \right) l(s,\ t)ds$$
$$= \int_a^b h(r,\ z) \left(\int_a^b k(z,\ s)l(s,\ t)ds \right) dz$$
$$= \int_a^b h(r,\ z)(kl)(z,\ t)dz$$
$$= h(kl)(r,\ t)$$

Another property of continuous kernels is their iteration property. That is, according to Eq. (4.23), we have:

$$k = h \Rightarrow \left\|k^2\right\|_c \le \|k\|_c^2$$

and by induction, it can be easily shown that

$$\left\|k^n\right\|_c \le \|k\|_c^n \tag{4.24}$$

4.3.2 Remark

If $k(s,t)$ is continuous and for each continuous function $x(t)$, we have:

$$\int_a^b k(s,t)x(t)dt = 0$$

Then, $k(s,t) = 0$. i.e.,

$$\forall x(kx = 0 \Rightarrow k = 0) \tag{4.25}$$

The above discussions can also be stated as follows.
If for each continuous function, we have

$$k_1 x = k_2 x \Rightarrow k_1 = k_2$$

by assuming $k = k_1 - k_2$, we obtain Eq. (4.25). To prove Eq. (4.25), suppose that at the kernel of the integral equation, s is arbitrary and constant hereafter. That is, $k(s,t) = f(t)$ is a function of t. Therefore, the complex conjugate of $k(s,t)$ will also be a function of t, and since $x(t)$ is an arbitrary function, then it can be assumed that $\overline{k(s,t)} = x(t)$. Therefore, we have:

$$\int_a^b k(s,t)\overline{k(s,t)}dt = \int_a^b |k(s,t)|^2 dt$$

If we assume

$$\int_a^b |k(s,t)|^2 dt = 0$$

According to the continuity of the positive function $|k(s,t)|^2$, we have

$$k(s,t) = 0$$

and because s is arbitrary, then

$$\forall t \forall s (k(s,\ t) = 0)$$

In fact, we proved the following lemma.

4.3.3 Lemma

$$\forall t \forall s (k(s, t)x(t) = 0 \Rightarrow k(s, t) = 0)$$

4.3.4 Definition

A kernel is called L^2 kernel and written as $k \in L^2(a, b)$, if we have

$$(1)\ \int_a^b \int_a^b |k(s, t)|^2 dt ds < \infty$$

$$(2)\ \forall s \int_a^b |k(s, t)|^2 dt < \infty,\ \text{is measurable}$$

$$(3)\ \forall t \int_a^b |k(s, t)|^2 ds < \infty,\ \text{is measurable}$$

We now state the Fubini, Tonley and Hopson theorems without proof.

4.3.5 Fubini Theorem

If $\int_a^b \int_a^b f(s, t) ds dt$ exists in the Lebesgue sense, then

$$\int_a^b f(s, t) dt$$

is almost ubiquitous. That means

$$m\left(\left\{s | \int_a^b f(s, t) dt \text{ does not exist}\right\}\right) = 0$$

and is also integrable with respect to s. That means

$$\int_a^b \left(\int_a^b f(s,t)dt \right) ds < \infty$$

and

$$\begin{aligned} \int_a^b \int_a^b f(s,t)dsdt &= \int_a^b ds \int_a^b f(s,t)dt \\ &= \int_a^b dt \int_a^b f(s,t)ds \end{aligned}$$

4.3.6 Tonley–Hopson Theorem

If $f(s,t)$ is measurable and one of the following integrals is available

$$\int\int |f(s,\ t)| dsdt \text{ or } \int ds \int |f(s,\ t)| dt \text{ or } \int dt \int |f(s,\ t)| ds$$

Then the following integrals exist and are equal.

$$\int\int |f(s,t)| dsdt = \int ds \int |f(s,t)| dt = \int dt \int |f(s,t)| ds$$

4.3.7 Lemma

If $x,\ k \in L^2(a,b)$ and

$$y(s) = \int_a^b k(s,\ t)x(t)dt$$

Then

$$y \in L^2(a,\ b)$$

Proof Suppose that s is arbitrary and constant hereafter. We have:

$$\left| \int_a^b f(t)g(t)dt \right|^2 \le \|f\|^2 \cdot \|g\|^2$$

So according to the Cauchy–Schwarz equation, we have:

$$|y(s)|^2 = \left| \int_a^b k(s, t)x(t)dt \right|^2 \leq \left(\int_a^b |k(s, t)|^2 dt \right) \left(\int_a^b |x(t)|^2 \right)$$

where $\int_a^b |k(s, t)|^2 dt \in L^2$, then

$$\int_a^b |y(s)|^2 ds \leq \left(\int_a^b \int_a^b |k(s, t)|^2 dt ds \right) \int_a^b |x(t)|^2 dt < \infty$$

Because we work in L^2, then $y(s)$ is measurable. Therefore, it can be said that:

$$\|y\|_2 \leq \|k\|_2 \cdot \|x\|_2$$

where

$$\|k\|_2 = \left(\int_a^b \int_a^b |k(s, t)|^2 dt ds \right)^{\frac{1}{2}}, \ \|x\|_2 = \left(\int_a^b |x(t)|^2 dt \right)^{\frac{1}{2}}$$

4.4 Fredholm Integral Equation of the Second Type

In this section, we want to study the Fredholm integral equation of the second type in the matrix form. We discuss the availability and uniqueness of the answer, its adjoint equations, regular values and other properties.

$$x(s) = y(s) + \lambda \int_a^b k(s, t)x(t)dt, \ \lambda \neq 0 \tag{4.26}$$

or

$$x = y + \lambda kx, \ \lambda \neq 0$$

therefore

$$x - \lambda kx = y$$

then

$$(I - \lambda k)x = y$$

As a result, assuming that matrix $I - \lambda k$ is reversible, we have

$$x = (I - \lambda k)^{-1} y$$

We know that $(I - \lambda k)^{-1}$ is itself a matrix. In order to write the answer of the integral equation as another integral equation, suppose

$$x = (I - \lambda k)^{-1} y := (I + \lambda h)y$$

so, we have the following two equations

$$(I - \lambda k)x = y \tag{4.27}$$

and

$$x = (I + \lambda h)y \tag{4.28}$$

From Eqs. (4.27) and (4.28), it can be written

$$(I - \lambda k)(I + \lambda h)y = y$$

so

$$(I - \lambda k)(I + \lambda h) = I$$

thus

$$I - \lambda k + \lambda h - \lambda^2 kh = I$$

Then, we have

$$h - k = \lambda kh$$

and also, if we set a value for y, we have

$$x = (I + \lambda h)(I - \lambda k)x$$

that means

$$I = (I + \lambda h)(I - \lambda k) = I + \lambda h - \lambda k - \lambda^2 kh$$

Then

$$h - k = \lambda hk \tag{4.29}$$

As a result, provided that the solution of the Fredholm integral equation of the second type exists, we have:

$$h - k = \lambda kh = \lambda hk$$

So, if Eq. (4.26) has an answer, it satisfies the following definition.

4.4.1 Definition—Regular Value

If for a value of λ, a kernel $h_\lambda \in L^2(a, b)$ exists such that

$$h_\lambda - k = \lambda h_\lambda k = \lambda k h_\lambda \tag{4.30}$$

then λ is called a regular value of the kernel k, h_λ is called corresponding solver kernel of λ, and Eq. (4.32) is called regular value of k.

4.4.2 Theorem

The solver kernel is unique.

Proof Suppose that we have at least two distinct solver kernels, such as h_1 and h_2, both hold in the solver equation. That means

$$h_1 - k = \lambda h_1 k = \lambda k h_1$$
$$h_2 - k = \lambda h_2 k = \lambda k h_2$$

By taking their difference, we have

$$h_1 - h_2 = \lambda(h_1 - h_2)k = \lambda k(h_1 - h_2)$$

So, by assuming $h = h_1 - h_2$ we will have

$$h = \lambda hk = \lambda kh$$

By forming hh_1, we have:

$$\begin{aligned} hh_1 &= h(k + \lambda kh_1) \\ &= hk + \lambda hkh_1 \\ &= hk + hh_1 \end{aligned}$$

Then

$$hk = 0 \;\&\; h = \lambda hk = 0$$

So,

$$h = h_1 - h_2 = 0$$

and this is a contradiction.

So, we proved that if λ is a regular value, then its corresponding solver kernel is unique.

4.4.3 Theorem

If h is the corresponding solver kernel of λ, then there is a unique x that satisfies in the following equation

$$x = y + \lambda kx \tag{4.31}$$

and is defined as follows.

$$x = y + \lambda hy \tag{4.32}$$

Proof First, we show that it holds in the following equation:

$$\begin{aligned} y + \lambda kx &= y + \lambda k(y + \lambda hy) \\ &= y + \lambda ky + \lambda^2 khy \end{aligned}$$

Also, since $h - k = \lambda kh = \lambda hk$, then:

$$\begin{aligned} y + \lambda kx &= y + \lambda ky + \lambda(h - k)y \\ &= y + \lambda ky + \lambda hy - \lambda ky \\ &= y + \lambda hy \\ &= x \end{aligned}$$

So, we show that x holds in Eq. (4.31); i.e., it is the answer. Now we show that if Eq. (4.31) holds, x is in the form of Eq. (4.32).

$$\begin{aligned} y + \lambda hy &= x - \lambda hx + \lambda hy \\ &= x - \lambda hx + \lambda h(x - \lambda hx) \\ &= x - \lambda hx + \lambda hx - \lambda(\lambda hk)x \\ &= x + \lambda(h - k - \lambda hk)x \\ &= x \end{aligned}$$

Now, since x holds in Eq. (4.31), and Eq. (4.32) must be hold, it is unique.

4.5 Continuous Kernels

Suppose that the kernel k is continuous. We want to prove that h_λ is continuous. We have:

$$h_\lambda - k = \lambda h_\lambda k = \lambda k h_\lambda$$

So

$$\begin{aligned} h_\lambda &= k + \lambda k h_\lambda \\ &= k + \lambda k(k + \lambda h_\lambda k) \\ &= k + \lambda k^2 + \lambda^2 k h_\lambda k \end{aligned}$$

$k + \lambda k^2$ is a combination of two continuous functions, so it suffices to prove that $k + \lambda^2 k h_\lambda k$ is continuous. For this purpose, we prove the following theorem:

4.5.1 Theorem

Suppose that $k, h, l \in L^2$, then:

$$|khl(s,\ t)| \leq \|h\| \ \cdot\ k(s)\ \cdot\ e(t)$$

where

$$\begin{aligned} k(s) &= \left(\textstyle\int |k(s,t)|^2 dt\right)^{\frac{1}{2}} \\ e(s) &= \left(\textstyle\int |l(s,t)|^2 dt\right)^{\frac{1}{2}} \end{aligned}$$

Proof We have

$$|t\langle hx,\ y\rangle| \leq \|hx\| \cdot \|y\| \leq \|h\| \cdot \|x\| \cdot \|y\|$$

Suppose that s and t are arbitrary and hereafter constant. We define:

$$x(v) = l(v, t), y(u) = \overline{k(s, u)}$$

therefore

$$\begin{aligned}(hx)(u) &= \int_a^b h(u,\ v)x(v)dv \\ &= \int_a^b h(u,\ v)l(v,\ t)dv\end{aligned}$$

then

$$\begin{aligned}\langle hx,\ y\rangle &= \int_a^b (hx)(u)\overline{y(u)}du \\ &= \int_a^b (hx)(u)k(s,\ u)du \\ &= \int_a^b\int_a^b h(u,\ v)l(v,\ t)k(s,\ u)dudv \\ &= \int_a^b\left(\int_a^b h(u,\ v)h(v,\ t)du\right)k(s,\ u)dv \\ &= khl(s,\ t)\end{aligned}$$

So, we showed that $\langle hx,\ y\rangle = khl(s,\ t)$

$$\begin{aligned}\|x\| &= \left(\int_a^b |x(v)|^2 dv\right)^{\frac{1}{2}} \\ &= \left(\int_a^b |l(v, t)|^2 dv\right)^{\frac{1}{2}} \\ &= e(t)\end{aligned}$$

and

$$\begin{aligned}\|y\| &= \left(\int |y(u)|^2 du\right)^{\frac{1}{2}} \\ &= \left(\int \overline{|k(s, u)|}^2 du\right)^{\frac{1}{2}}\end{aligned}$$

$$= \left(\int |k(s,u)|^2 du\right)^{\frac{1}{2}}$$
$$= k(s)$$

The proof is completed.

Now, we prove that $kh_\lambda k$ is continuous. Because k is continuous on a closed interval, so it has a uniform continuity. That means:

$$\forall \varepsilon \exists \delta \forall s \forall s' \forall t\ \forall t'$$
$$\left(\varepsilon > 0, \delta > 0, \sqrt{(s-s')^2 + (t-t')^2} < \delta \Rightarrow \left|k(s,t) - k\left(s', t'\right)\right| < \varepsilon\right)$$

Suppose ε' is arbitrary and constant hereafter. We have:

$$\varepsilon = \frac{\varepsilon'}{2k_1 \|h_\lambda\| \sqrt{b-a}}$$

where

$$\forall s (|k(s)| \leq k_1)$$

Because $k(s)$, for every t, is constant and continuous, on a closed and bounded interval, it takes its limit. Given that:

$$\left|t - t'\right| \leq \sqrt{(s-s')^2 + (t-t')^2}$$
$$\left|s - s'\right| \leq \sqrt{(s-s')^2 + (t-t')^2}$$

therefore

$$\begin{aligned}
\left|kh_\lambda k(s,\ t) - kh_\lambda k\left(s',\ t'\right)\right| &\leq \left|kh_\lambda k(s,\ t) - kh_\lambda k\left(s,\ t'\right)\right| \\
&+ \left|kh_\lambda k\left(s,\ t'\right) - kh_\lambda k\left(s',\ t'\right)\right| \\
&= \left|kh_\lambda\left(k(s,\ t) - k\left(s,\ t'\right)\right)\right| \\
&+ \left|kh_\lambda\left(k\left(s,\ t'\right) - k\left(s',\ t'\right)\right)\right| \\
&\leq \|h_\lambda\| k(s) \left(\int_a^b \left|k(s,\ t) - k\left(s,\ t'\right)\right|^2 ds\right)^{\frac{1}{2}} \\
&+ \|h_\lambda\| k(s) \left(\int_a^b \left|k(s,\ t) - k\left(s',\ t'\right)\right|^2 ds\right)^{\frac{1}{2}} \\
&\leq \|h_\lambda\| k_1 \left(\frac{\varepsilon'}{2k_1 \|h_\lambda\|} \frac{\varepsilon'}{2k_1 \|h_\lambda\|}\right) \\
&= \varepsilon'
\end{aligned}$$

Therefore, $kh_{\lambda}k$ is continuous.

4.6 Adjoint Kernels

Sometimes, the work field is complex numbers, so according to the needs of the following subjects, we discuss adjoint kernels.

4.6.1 Definition

k^* is called the adjoint kernel of k, if

$$k^*(s,t) = \overline{k(s,t)}$$

For adjoint kernels, we have the following properties

(1) $(k^*)^* = k$
(2) $(k+l)^* = k^* + l^*$
(3) $(\lambda k)^* = \overline{\lambda} k^*$

If $k = k^*$, then k is called **Hermitian** or self-adjoint, and if we have $k = -k^*$, in this case k is called **skew-symmetric**.

To prove the property (2), we have:

$$(k+l)^*(s,t) = \overline{(k+l)(t,s)} = \overline{k(t,s)} + \overline{l(t,s)} = k^*(s,t) + l^*(s,t)$$

It can be shown that if k is a real and symmetric function, then it is hermitian. That means:

$$k(s,t) = k(t,s) = \overline{k(t,s)} = k^*(s,t)$$

So, $k = k^*$.

After mentioning and reviewing the above properties, we will examine other properties of the adjoint kernels.

4.6.2 The Combination of Two Adjoint Functions

$$(kl)^*(s,\ t) = \overline{(kl)(t,\ s)}$$

$$= \int_a^b \overline{k(t,\ u)l(u,\ s)}du$$

$$
\begin{aligned}
&= \int_a^b \overline{k(t,\ u)} \cdot \overline{l(u,\ s)}du \\
&= \int_a^b l^*(s,\ u)k^*(u,\ t)du \\
&= l^*k^*(s,\ t)
\end{aligned}
$$

Then, $(kl)^* = l^*k^*$.
It can also be shown that $\langle k^*x,\ y\rangle = \langle x,\ ky\rangle$.

$$
\begin{aligned}
\langle k^*x,\ y\rangle &= \int_a^b \left(\int_a^b k^*(s,\ t)x(t)dt \right) \overline{y(s)}ds \\
&= \int_a^b \left(\int_a^b \overline{k(t,\ s)}x(t)dt \right) \overline{y(s)}ds \\
&= \int_a^b x(t) \left(\int_a^b \overline{k(t,\ s)y(s)ds} \right) dt \\
&= \langle x,\ ky\rangle
\end{aligned}
$$

Therefore, according to this property, it can be written

$$
\langle kx, y\rangle = \langle (k^*)^* x, y\rangle = \langle x, k^* y\rangle
$$

4.6.3 Definition—Normal Kernel

If $kk^* = k^*k$, then k is called normal.

4.6.4 Remark

If k is Normal, then $kk^* = k^*k = k^2$. So, if k is Hermitian, it is definitely normal, but the opposite is not true.

4.6.5 Adjoint Equations

Adjoint equations can be introduced as with adjoint kernels. For the equation $x = y + \lambda kx$, the equation $u = v + \overline{\lambda}k^*u$ can also be considered as an adjoint equation, so that if h is a solver kernel of k corresponding to λ, then h^* is the solver kernel of k^* corresponding to $\overline{\lambda}$. Because

$$h - k = \lambda kh = \lambda hk$$

Therefore,

$$h^* - k^* = \overline{\lambda} h^* k^* = \overline{\lambda} k^* h^*$$

That means that h^* is solver kernel of k^* corresponding to $\overline{\lambda}$.
Consider the following integral equation:

$$x(s) = \lambda \int_a^b k(s, t)x(t)dt \tag{4.33}$$

We know that the function $x(t) = 0$ holds in this equation. Now if $x \neq 0$ and $x \in L^2(a, b)$ is the answer to the above integral equation, then λ is called a characteristic value of k and x is called a characteristic function corresponding to λ. The operator form of this equation is as $x = \lambda kx$.

It can be shown that any linear combination of the solutions of the above equation (characteristic functions corresponding to λ) is itself a characteristic function corresponding to λ. That is, if x_1 and x_2 are characteristic functions corresponding to λ, then:

$$x_1 = \lambda kx_1$$
$$x_2 = \lambda kx_2$$

therefore

$$\alpha_1 x_1 = \lambda k(\alpha_1 x_1)$$
$$\alpha_2 x_2 = \lambda k(\alpha_2 x_2)$$

as a result

$$0 \neq \alpha_1 x_1 + \alpha_2 x_2 = \lambda k(\alpha_1 x_1 + \alpha_2 k_2)$$

So $\alpha_1 x_1 + \alpha_2 x_2$ is also a characteristic function corresponding to λ.

4.6.6 Definition

The Set

$$\{x | x = \lambda kx, x \in L^2\}$$

is a subspace of L^2, which is called the characteristic subspace of k corresponding to λ.

4.6.7 Lemma

If x is a characteristic and continuous function, then $x \neq 0$.

Proof Suppose by contradiction that $\|x\| = 0$. Because $x \in L^2$, then x is equal to zero almost everywhere. Therefore,

$$\int_a^b k(s,t)x(t)dt = 0$$

So, according to Eq. (4.33):

$$\forall s(x(s) = 0)$$

That means $x \equiv 0$, which is a contradiction because the characteristic function is always nonzero.

4.6.8 Remark

If $x = \lambda kx$ and $\lambda = 0$, then λ is not a characteristic value of x. That means that if λ is the characteristic value, then, definitely $\lambda \neq 0$.

μ is the characteristic value or eigenvalue of the kernel k and $kx = \lambda^{-1}x = \mu x$ so that if $\mu = 0$, then $kx = 0$.

4.6.9 Theorem

If λ is a regular value of kernel k, then λ is not its characteristic value.

Proof Suppose that λ is the characteristic value, then $x \neq 0$ and $x = \lambda kx$. On the other hand, because λ is a regular value for k, so if $y = 0$, then $x = 0 + \lambda kx$ has an answer that is in the form of

$$x = y + \lambda hy = 0 + \lambda h0 = 0$$

which is a contradiction.

4.6.10 Theorem

If $x(s)$ is the characteristic function of the kernel $k(s,t) \in c[a,b] \times [a,b]$, then x is continuous.

Proof We have,

$$x(s) = \lambda \int_a^b k(s,\ t)x(t)dt, \quad x \in L^2$$

we prove that:

$$\forall \varepsilon \exists \delta \forall s \forall s' \left(|s - s'| < \delta \Rightarrow |x(s) - x(s')| < \varepsilon \right)$$

therefore

$$\begin{aligned} x(s) - x(s') &= \lambda \int_a^b k(s,\ t)x(t)dt - \lambda \int_a^b k(s',\ t)x(t)dt \\ &= \lambda \int_a^b \left(k(s,\ t) - k(s',\ t)\right)x(t)dt \end{aligned}$$

as a result

$$\begin{aligned} |x(s) - x(s')| &\le |\lambda| \cdot \int_a^b |k(s,\ t) - k(s',\ t)||x(t)|dt \\ &\le |\lambda| \cdot x \left(\int_a^b |k(s,\ t) - k(s',\ t)|^2 dt \right)^{\frac{1}{2}} \end{aligned}$$

Due to the continuity of the kernel k, for every arbitrary ε', there is a δ that for every s,

$$\sqrt{(s - s')^2 + (t - t')^2} < \delta$$

then

$$|k(s,t) - k(s',t)| < \varepsilon'$$

therefore,

$$|x(s) - x(s')| < |\lambda| \cdot \|x\| \left((b - a)\varepsilon'^2\right)^{\frac{1}{2}}$$

it suffices to assume ε as follows.

$$\varepsilon' = \frac{\varepsilon}{\sqrt{b-a}(1 + |\lambda| \cdot \|x\|)}$$

as a result

$$|x(s) - x(s')| \leq |\lambda| \cdot \|x\| \cdot \frac{\varepsilon}{1 + |\lambda| \cdot \|x\|} < \varepsilon$$

4.6.11 Example

Suppose that k is continuous and $x \in L^2$ is a discontinuous function and the characteristic function:

$$k(s,t) = 2s - 5st, \; x(t) = \begin{cases} t^{-\frac{1}{3}}, & t \neq 0, t \in [0,1] \\ 0, & t = 0 \end{cases}$$

In this case,

$$\begin{aligned} \int_0^1 k(s, t)x(t)dt &= \int_0^1 (2s - 5st)t^{-\frac{1}{3}}dt \\ &= 2s\int_0^1 t^{-\frac{1}{3}}dt - 5s\int_0^1 t^{\frac{2}{3}}dt = 0, \quad t \neq 0 \end{aligned}$$

So, we can say that $x(t)$ for $t \neq 0$ is a characteristic function corresponding to $\lambda = 0$, i.e., $kx = \lambda x$:

Now, we express the concepts of different convergences.

4.6.12 Definition

Suppose that $\{x_n(s)\}$ is a sequence of functions L^2, i.e.,

$$\forall n(x_n(s) \in L^2(a,b))$$

Sequence that $\{x_n\}$ has a relatively uniformly convergence to x if and only if

$$\begin{aligned} &\exists p(s)\forall\varepsilon\exists N\forall n\forall s \\ &(p(s) \geq 0 \;\&\; p(s) \in L^2 \;\&\; n \geq N \Rightarrow |x_n(s) - x(s)| \leq \varepsilon p(s)) \end{aligned}$$

4.6.13 Definition—Point Wise Convergence

We say that the sequence $\{x_n\}$ convergence to x point wisely, if and only if

$$\forall\varepsilon\forall n\forall s\exists N(s,\ \varepsilon)(\varepsilon > 0\ \&\ n \geq N \Rightarrow |x_n(s) - x(s)| < \varepsilon)$$

That means that convergence depends on the points.

4.6.14 Definition—Uniformly Convergence

We say that sequence $\{x_n\}$ has a uniformly convergence to x if and only if

$$\forall\varepsilon\exists N(\varepsilon)\forall n\forall s(\varepsilon > 0\ \&\ n \geq N \Rightarrow |x_n(s) - x(s)| \leq \varepsilon)$$

According to these definitions, it can be concluded that:

Uniformly convergence $\Rightarrow$ Relatively uniformly convergence $\Rightarrow$ Pointwise convergence.

To prove the first part, it suffices to assume that $0 < p(s) = 1 \in L^2(a, b)$.

To prove the second part, we do the following

$$\forall\varepsilon\exists N\forall n\forall s\left(n \geq N\ \&\ \varepsilon = \frac{\varepsilon'}{1 + p(s)} \Rightarrow |x_n(s) - x(s)| \leq \varepsilon'\right)$$

We can also consider N as dependent on ε and s, in this case, we have a pointwise convergence.

4.6.15 Example

Suppose

$$x_n(t) = \begin{cases} \frac{1}{nt^{\frac{1}{3}}}, & t \neq 0,\ t \in [0, 1] \\ 0, & t = 0 \end{cases}$$

$$p(t) = \begin{cases} \frac{1}{t^{\frac{1}{3}}}, & t \neq 0 \\ 0, & t = 0 \end{cases}$$

$$x(s) = 0, \quad s \in [0, 1]$$

Because $p(t) \geq 0$ and

$$\int_0^1 p^2(s)ds = \int_0^1 s^{-\frac{2}{3}}ds = 3 > 0$$

Therefore, $p(t) \in L^2(0, 1)$ and also

$$\int_0^1 x_n^2(s)ds = \frac{1}{n}\int_0^1 s^{-\frac{2}{3}}ds = \frac{3}{n}$$

So, $x_n(t) \in L^2(0, 1)$. Also:

$$|x_n(s) - x(s)| = |x_n(s)| = \frac{1}{n}p(s)$$

we have

$$\forall\varepsilon\exists N\forall n\left(n \geq N \Rightarrow \frac{1}{n} < \varepsilon\right)$$

then

$$\forall\varepsilon\exists N\forall n\forall s\left(n \geq N \Rightarrow |x_n(s) - x(s)| = \frac{1}{n}p(s) \leq \varepsilon' p(s)\right)$$

which has relatively uniformly convergence but does not have uniformly convergence, because to prove it, we must prove the following formula:

$$\forall\varepsilon\exists N\forall n\forall s\left(n \geq N \Rightarrow \frac{1}{nt^{\frac{1}{3}}} < \varepsilon'\right)$$

Suppose that we have uniformly convergence, so we can assume that $\varepsilon = 1$ and $s = \frac{1}{N^3} \in [0, 1]$; therefore,

$$\frac{1}{nt^{\frac{1}{3}}} = \frac{1}{N \cdot \frac{1}{N}} = 1 < 1$$

which is a contradiction.

If the convergence is relatively uniform, it may not be uniform.

Relatively uniformly convergence can also be expressed by the following definition.

4.6.16 Definition

A necessary and sufficient condition for $x_n(s)$ to have an uniform relative convergence to $x(s)$ is that

$$\forall\varepsilon\exists n_0\forall n\forall m\forall s$$
$$(n \geq n_0 \ \& \ m \geq n_0 \ \& \ s \in [a, b] \Rightarrow |x_n(s) - x(s)| \leq \varepsilon p(s))$$

We now consider the convergence discussed the sequences for series.

4.6.17 Definition

Series $\sum_{i=1}^{\infty} x_i(s)$ converges relatively uniformly to $x(s)$ if and only if series $\sum_{i=1}^{n} x_i(s)$ converges relatively uniformly to $x(s)$.

The same convergences hold for kernels.

4.6.18 Definition

Sequence $\{k_n(s,\ t)_n\}$ has a relatively uniformly convergence to $k(s,\ t)$ if

$$\exists p(s,\ t)\forall\varepsilon\exists n_0\forall n\forall(s,\ t)$$
$$((n \geq n_0 \ \& \ (s,\ t) \in [a,\ b] \times [a,\ b] \ \& \ p(s,\ t) \geq 0$$
$$\& \ p(s,\ t) \in L^2) \Rightarrow |k_n(s,\ t) - k(s,\ t)| < \varepsilon p(s,\ t))$$

4.6.19 Theorem

If $x_n(s)$ has a relatively uniformly convergence to $x(s)$ and $y(s) \in L^2$, then $\langle x_n(s), y(s)\rangle$ converges to $\langle x,\ y\rangle$.

That means:

$$\lim_{n\to\infty} \int_a^b x_n(s)\overline{y(s)}ds = \int_a^b x(s)\overline{y(s)}ds$$

Proof Given that $x_n(s)$ has a relatively uniformly convergence to $x(s)$, therefore:

$$|x_n(s) - x(s)| < \varepsilon' p(s)$$

where $\varepsilon' = \frac{\varepsilon}{1+l}$ and $0 \le l = \int_a^b p(s)|y(s)|ds$. As the result, we can write:

$$\begin{aligned} |\langle x_n(s),\ y(s)\rangle - \langle x(s),\ y(s)\rangle| &= \left|\int_a^b \left(x_n(s) - x(s)\overline{y(s)}ds\right)\right| \\ &\le \int_a^b |x_n(s) - x(s)||y(s)|ds \\ &< \varepsilon' \int_a^b p(s)|y(s)|ds \\ &= \frac{\varepsilon l}{1+l} \\ &< \varepsilon \end{aligned}$$

According to Theorem (4.6.19), it can be concluded that if $x_n(s)$ has a relatively uniformly convergence to $x(s)$, then $\langle x_n(s),\ x_n(s)\rangle$ has a relatively uniformly convergence to $\langle x(s),\ x(s)\rangle$, or in other words, $\|x_n(s)\|$ has a relatively uniformly convergence to $\|x(s)\|$.

4.6.20 Theorem

If $x \in L^2$ and $k_n(s,t)$has a relatively uniformly convergence to $k(s,t)$, then $\int_a^b k_n(s,t)x(t)dt$has a relatively uniformly convergence to $\int_a^b k(s,t)x(t)dt$.

Proof Because $k_n(s,t)$ has a relatively uniformly convergence to $k(s,t)$, so according to Definition (4.6.18),

$$|k_n(s,t) - k(s,t)| < \varepsilon' p(s,t)$$

as a result

$$\begin{aligned} &\left|\int_a^b k_n(s,t)x(t)dt - \int_a^b k(s,t)x(t)dt\right| \\ &\le \int_a^b |k_n(s,t) - k(s,t)||x(t)|dt \\ &\le \varepsilon' \int_a^b p(s,\ t)|x(t)|dt \\ &= \varepsilon' q(s) \end{aligned}$$

where $q(s) \geq, 0$ $q \in L^2$ and $\varepsilon' = \frac{\varepsilon}{1+q(s)}$.

The following theorem points out the relatively uniformly convergence of a sequence of productions of finite kernels.

4.6.21 Theorem

If $l \in L^2$ and $k_n(s,t)$ have a relatively uniformly convergence to $k(s,t)$, then $k_n l$ has a relatively uniformly convergence to kl.

Proof Because $k_n(s,t)$ has a relatively uniformly convergence to $k(s,t)$, so according to the definition of (4.6.18),

$$|k_n(s,u) - k(s,u)| < \varepsilon' p(s,u)$$

as a result

$$\begin{aligned}
&\left|\int_a^b k_n(s,u)l(u,t)du - \int_a^b k(s,u)l(u,t)du\right| \\
&\leq \int_a^b |k_n(s,u) - k(s,u)||l(u,t)|du \\
&\leq \varepsilon' \int_a^b p(s,u)|l(u,t)|du \\
&= \varepsilon' q(s,t)
\end{aligned}$$

where $q(s,t) \geq 0$, $q \in L^2$ and $\varepsilon' = \frac{\varepsilon}{1+q(s,t)}$.

4.6.22 Theorem

If the assumptions of Theorem (4.6.21) hold, then lk_n has a relatively uniformly convergence to lk.

Proof The burden of the proof is the responsibility of the readers.

The above-mentioned theorems can be stated about the series. For example, if $\sum_{i=1}^{\infty} x_i(s)$ has a relatively uniformly convergence to $x(s)$, and if we assume $y_n(s) = \sum_{i=1}^{n} x_i(s)$, then $y_n(s)$ has a relatively uniformly convergence to $y(s)$. So

$$\forall \varepsilon \exists n_0 \forall n \forall m \forall s$$
$$(n \geq n_0 \ \& \ m \geq n_0 \ \& \ s \in [a, b] \Rightarrow |y_n(s) - y_m(s)| < \varepsilon p(s))$$

Therefore, because the series is convergent, we have:

$$\forall \varepsilon \exists n_0 \forall n \forall m \forall s$$
$$\left(n > m \geq n_0 \ \& \ s \in [a, b] \Rightarrow \left| \sum_{i=m+1}^{n} x_i(s) \right| < \varepsilon p(s)\right)$$

4.6.23 Theorem

If $\sum_{i=1}^{\infty} x_i(s)$ has a relatively uniformly convergence to $x(s)$ and $y \in L^2$, then $\sum_{i=1}^{\infty} \langle x_i(s), y(s) \rangle$ converges to $\langle x(s), y(s) \rangle$.

That means:

$$\sum_{i=1}^{\infty} \int_a^b x_i(s)\overline{y(s)}ds = \int_a^b \left(\sum_{i=1}^{\infty} x_i(s) \right) \overline{y(s)}ds$$

is convergent to $\int_a^b x(s)\overline{y(s)}ds$.

Proof Because we have a relatively uniformly convergence, then

$$\forall \varepsilon' \exists n_0 \forall n \forall m \forall s$$
$$\left(n > m \geq n_0 \ \& \ s \in [a, b] \Rightarrow \left| \sum_{i=m+1}^{n} x_i(s) \right| < \varepsilon' p(s) \right)$$

We prove that

$$\forall \varepsilon \exists n_0 \forall n \forall m \forall s (n > m \geq n_0 \ \& \ s \in [a, b]$$
$$\Rightarrow \left| \sum_{i=m+1}^{n} \langle x_i(s), y(s) \rangle \right| < \varepsilon)$$

For this purpose,

$$\left|\sum_{i=m+1}^{n}\langle x_i(s), y(s)\rangle\right| = \left|\int_a^b \sum_{i=m+1}^{n} x_i(s)\overline{y(s)}ds\right| =$$
$$\leq \int_a^b \left|\sum_{i=m+1}^{n} x_i(s)\right| |y(s)|ds$$
$$\leq \varepsilon' \int_a^b p(s)|y(s)|ds$$

Suppose $l = \int_a^b p(s)|y(s)|ds$, then

$$\left|\sum_{i=m+1}^{n}\langle x_i(s),\ y(s)\rangle\right| < \varepsilon' l$$

It suffices to assume $\varepsilon' = \frac{\varepsilon}{1+l}$ that in this case, the sentence is true.

4.6.24 Theorem

If $\sum_{n=1}^{\infty} k_n(s,t)$ has a relatively uniformly convergence to $k(s,t)$ and $x \in L^2$, then $\sum_{n=1}^{\infty}\int_a^b k_n(s,t)x(t)dt$ has a relatively uniformly convergence to $\int_a^b k(s,t)x(t)dt$.

Proof Because $\sum_{n=1}^{\infty} k_n(s,t)$ has a relatively uniformly convergence to $k(s,t)$, then

$$\forall\varepsilon\exists n_0\forall n\forall m\forall s\forall t$$
$$\left(n > m \geq n_0 \ \&\ s,t \in [a,b] \Rightarrow \left|\sum_{i=m+1}^{n} k_i(s,t)\right| < \varepsilon p(s,t)\right)$$

where $p(s,t) \geq 0, p \in L^2$. According to what was mentioned before, we have:

$$\left|\sum_{i=m+1}^{n}\int_a^b k_i(s,t)x(t)dt\right| = \left|\int_a^b \sum_{i=m+1}^{n} k_i(s,t)x(t)dt\right| =$$
$$\leq \int_a^b\left|\sum_{i=m+1}^{n} k_i(s,t)\right||x(t)|dt$$
$$\leq \varepsilon \int_a^b p(s,t)|x(t)|dt$$
$$= \varepsilon q(s)$$

where $q(s) \geq 0, q \in L^2$, and as a result, the sentence is true.

4.6.25 Theorem

If $\sum_{n=1}^{\infty} k_n(s,\ t)$ has a relatively uniformly convergence to $k(s,\ t)$ and $l \in L^2$, then $\sum_{n=1}^{\infty} k_n l$ has a relatively uniformly convergence to kl and $\sum_{n=1}^{\infty} lk_n$ has a relatively uniformly convergence to lk.

Proof Given that

$$k_n l = \int_a^b k_n(s,\ u)l(u,\ t)du$$

We must prove that $\sum_{n=1}^{\infty}\int_a^b k_n(s,\ u)l(s,\ t)du$ has a relatively uniformly convergence to $\int_a^b k(s,\ u)l(s,\ t)du$. First, suppose that $\varepsilon > 0$ is arbitrary where:

$$\exists n_0 \forall n \forall m \forall s \forall t$$
$$\left(n > m \geq n_0 \ \&\ s, u \in [a,\ b] \Rightarrow \left|\sum_{i=m+1}^{n} k_i(s,\ u)\right| < \varepsilon p(s,\ u)\right)$$

where $p(s,\ u) \geq 0$, $p \in L^2$. Therefore,

$$\exists n_0 \forall n \forall m \forall s \forall t (n > m \geq n_0 \ \&\ s, u \in [a,\ b]$$
$$\Rightarrow \left|\sum_{i=m+1}^{n}\int_a^b k_i(s,\ u)l(u,\ t)du\right| < \varepsilon q(s,\ u))$$

where $q(s,u) \geq 0$, $q \in L^2$. Because

$$\left|\sum_{i=m+1}^{n} \int_a^b k_i(s, u) l(u, t) du\right| \leq \int_a^b \left|\sum_{i=m+1}^{n} k_i(s, t)\right| |l(u, t)| du$$
$$< \varepsilon \int_a^b p(s, t) |l(u, t)| du$$
$$= \varepsilon q(s, t)$$

Chapter 5
Numerical Solution of Integral Equations

5.1 Introduction

In a lot of applications of mathematics, finding the analytical solution is too complicated; in recent years, a lot of attention has been devoted by researchers to find the numerical solution of equations. In this chapter, we are going to look at several numerical methods, and at the end, we will discuss the second kind integral equations.

5.2 Neumann Series

In this section, we want to introduce the solution of a Fredholm integral equation in the matrix form as a series called the Neumann series. In the meantime, the solver kernel h_λ can also be displayed as a series. Obviously, by applying some conditions to the series sentences, it can be claimed that the integral equation has a solution.

Suppose that

$$x(s) = y(s) + \lambda \int_a^b k(s,t)x(t)dt$$

That means $x = y + \lambda kx$. If λ is a regular value for the kernel k, then

$$\exists h_\lambda : h_\lambda - k = \lambda h_\lambda k = \lambda k h_\lambda$$

then

$$h_\lambda(I - \lambda k) = k \tag{5.1}$$

T. Allahviranloo and A. Esfandiari, *A Course on Integral Equations with Numerical Analysis*, Mathematical Engineering, https://doi.org/10.1007/978-3-030-85350-1_5

$$(I - \lambda k)h_\lambda = k \tag{5.2}$$

The Eqs. (5.1) and (5.2) are matrix equations so that the matrix $I - \lambda k$ is invertible, so according to the Eq. (5.1), it is concluded:

$$\begin{aligned} h_\lambda &= k(I - \lambda k)^{-1} = k\frac{1}{I - \lambda k} \\ &= k\big(I + \lambda k + (\lambda k)^2 + \cdots\big) = k + \lambda k^2 + \lambda^2 k^3 + \cdots \end{aligned}$$

And from Eq. (5.2), it can be written:

$$\begin{aligned} h_\lambda &= (I - \lambda k)^{-1}k = \frac{1}{I - \lambda k}k \\ &= \big(I + \lambda k + (\lambda k)^2 + \cdots\big)k = k + \lambda k^2 + \lambda^2 k^3 + \cdots \end{aligned}$$

That means that in both cases, we can see that h_λ is in terms of the above series. The determination of h_λ as the above series can be done in another way:

$$x = y + \lambda k x$$

The following iterative equation can be written as follows:

$$x_1 = y,\ x_{n+1} = y + \lambda k x_n$$

So

$$\begin{aligned} x_2 &= y + \lambda k y \\ x_2 &= y + \lambda k(y + \lambda k y) = y + \lambda k y + \lambda^2 k^2 y \\ &\vdots \end{aligned}$$

And finally

$$\begin{aligned} x_{n+1} &= y + \lambda k y + \lambda^2 k^2 y + \cdots + \lambda^n k^n y \\ &= y + \lambda\big(k + \lambda k^2 + \lambda^2 k^3 + \cdots + \lambda^{n-1} k^n\big)y \end{aligned}$$

Given that the equation $x = y + \lambda k x$ has a unique answer as $x = y + \lambda h_\lambda y$, then

$$x = y + \lambda\big(k + \lambda k^2 + \cdots\big)y$$

in this case, $h_\lambda = k + \lambda k^2 + \cdots$.

The following theorem discusses the relatively uniformly convergence of Neumann series.

5.2.1 Theorem

Suppose that λ *is a regular value of the kernel* $k(s,t)$*and* $|\lambda| \cdot \|k\| < 1$*, in which case, the following series*

$$k(s,t) + \lambda k^2(s,t) + \cdots + \lambda^{n-1}k^n(s,t) + \cdots$$

has a relatively uniformly convergence to $h_\lambda = (s,t)$.

Proof The general term of the series is as follows:

$$n \geq 2, \left|\lambda^{n-1}k^n(s,t)\right| = |\lambda|^{n-1} \cdot \left|k \cdot k^{n-2} \cdot k\right|$$

on the other hand, we had

$$|khl(s,t)| = \|h\| \cdot k(s) \cdot l(t)$$

where

$$k(s) = \left(\int_a^b |k(s,t)|^2 dt\right)^{\frac{1}{2}}, k(t) = \left(\int_a^b |l(s,t)|^2 dt\right)^{\frac{1}{2}}$$

then

$$\begin{aligned}\left|\lambda^{n-1}k^n(s,t)\right| &\leq |\lambda|^{n-1} \cdot \|k^{n-2}\| \cdot k(s) \cdot k(t)\\ &\leq |\lambda|^{n-1} \cdot \|k\|^{n-2} \cdot k(s) \cdot k(t)\end{aligned}$$

suppose

$$0 \leq p(s,t) = k(s) \cdot k(t) \in L^2$$

also

$$|\lambda|^{n-1} \cdot \|k\|^{n-2} = a_n$$

Therefore, according to the ratio test, we have:

$$\frac{a_{n+1}}{a_n} = \frac{|\lambda^n| \cdot \|k\|^{n-1}}{\left|\lambda^{n-1}\right| \cdot \|k\|^{n-2}} = |\lambda| \cdot \|k\| < 1$$

So $\sum_n a_n$ is convergent. Therefore, it can be concluded that $\sum_{n=1}^{\infty} \left|\lambda^{n-1} k^n(s,t)\right|$ has a uniformly convergence to $z(s,t)$. Therefore, it can be written:

$$\sum_{i=m+1}^{n} \left|\lambda^{i-1} k^i(s,t)\right| \leq \left(\sum_{i=m+1}^{n} a_i\right) p(s,t) < \varepsilon p(s,t)$$

So, we have a relatively uniformly convergence to $z(s,t)$.

Thus, the following series is convergent.

$$\begin{aligned} &k(s,t) + \lambda k^2(s,t) + \cdots + \lambda^{n-1} k^n(s,t) + \cdots \\ &= k(s,t) + \lambda k(s,t)\big(k(s,t) + \lambda k^2(s,t) + \cdots\big) \end{aligned}$$

as a result

$$z(s,t) = k(s,t) + \lambda k(s,t) z(s,t)$$

that means

$$z - k = \lambda kz,\ z - k = \lambda zk$$

and because h_λ is unique, so $z(s,t) = h_\lambda$.

The following example shows that the condition $|\lambda| \cdot \|k\| < 1$ is not a necessary condition.

5.2.2 *Example*

Suppose that $k(s,t) = u(s)\overline{v(t)}$ and

$$\langle u, v\rangle = \int_a^b u(t)\overline{v(t)}dt = 0$$

where $u(t) = \sin t$ and $v(t) = \cos t$.

$$k^2(s,t) = \int_a^b k(s,z)k(z,t)dz$$

$$= \int_a^b u(s)\overline{v(z)}u(z)\overline{v(t)}dz$$

$$= u(s)\overline{v(t)} \int_a^b u(z)\overline{v(z)}dz$$

$$= 0$$

then

$$k^2 = 0, \cdots k^n = 0, n \geq 2$$

That means that the series is zero from the second term onward and does not depend on λ and $h_\lambda = k$.

5.2.3 *Error Calculation*

We employ the Neumann method, which we examined in the Sect. (5.1), to solve the integral Eq. (5.1). We have:

$$x = y + \lambda ky + \lambda^2 k^2 y + \cdots$$
$$x_0 = y, x_{n+1} = y + \lambda k x_n$$

We define:

$$e_n = x_n - x$$

Given that $x = y + \lambda kx$, then

$$x_{n+1} - x = \lambda k(x_n - x)$$

therefore

$$e_{n+1} = \lambda k e_n$$

as a result

$$\|e_{n+1}\| \leq |\lambda| \cdot \|k\| \cdot \|e_n\|$$

on the other hand

$$e_{n+1} - e_n = (x_{n+1} - x) - (x_n - x) = x_{n+1} - x_n$$

so

$$e_n = e_{n+1} - (x_{n+1} - x_n)$$

then

$$\begin{aligned} \|e_n\| &\leq \|e_{n+1}\| + \|x_{n+1} - x_n\| \\ &\leq |\lambda| \cdot \|k\| \cdot \|e_n\| + \|x_{n+1} - x_n\| \end{aligned}$$

therefore

$$(1 - |\lambda| \cdot \|k\|) \cdot \|e_n\| \leq \|x_{n+1} - x_n\|$$

So, it can be said that

$$\|e_n\| \leq \frac{\|x_{n+1} - x_n\|}{1 - |\lambda| \cdot \|k\|}$$

is the posterior upper bound for the error of x_n.

For constant s,

$$\int_a^b k(s,t)x(t)dt \simeq \sum_{j=1}^{n} w_j k(s,t_j)x(t_j)$$

then

$$x_1(s) \simeq y(s) + \lambda \sum_{j=1}^{n} w_j k(s,t_j)y(t_j) \tag{5.3}$$

it can also be written

$$x_{2s}(s) \simeq y(s) + \lambda \sum_{j=1}^{n} w_j k(s,t_j)y(t_j) \tag{5.4}$$

And according to Eq. (5.3), we will have:

$$x_1(t_i) \simeq y(t_i) + \lambda \sum_{j=1}^{n} w_j k(t_i, t_j) y(t_j) \tag{5.5}$$

Which by substituting the Eq. (5.5) in Eq. (5.4), we have:

$$x_2(s) \simeq y(s) + \lambda^2 \sum_{j=1}^{n} \sum_{i=1}^{n} w_i w_j k(s, t_j) k(t_i, t_j) y(t_j)$$
$$+ \lambda \sum_{i=1}^{n} w_i k(s, t_i) y(t_i)$$

Similarly, to calculate $x_3(s)$, it is only necessary to calculate $x_2(t_i)$ for every i. In these calculations, we assume

$$[k(t_i, t_j)]_{i,j} = k_{ij},\ \|\lambda k_{ij}\| < 1$$

According to the above discussions, a sequence of x_n s is obtained which converges to the answer of the integral equation and the convergence condition is $\|\lambda k_{ij}\| < 1$. So we can say that the integral equation

$$x(s) = y(s) + \lambda \int_a^b k(s, t) x(t) dt$$

is solved using the following iterative method:

$$x_{n+1} = y(s) + \lambda \int_a^b k(s, t) x_n(t) dt$$
$$= y(s) + \lambda \sum_{j=1}^{n} w_j k(s, t_j) x_n(t_j),\ s = s_i, i = 1, \ldots, n \tag{5.6}$$

where

$$(x_n)_i = x_n(t_i),\ k_{ij} = w_j k(t_i, t_j),\ y_i = y(t_i)$$

then

$$x_{n+1} = y + \lambda k x_n$$
$$x_R = y + \lambda k x_R \tag{5.7}$$

The necessary condition is also $\|\lambda k\| < 1$.

It is possible that $\|\lambda k_{ij}\| < 1$, but $\|\lambda k\| \nless 1$ (k is the kernel of the integral equation).

Here x_R in Eq. (5.8) is an integration rule that depends on the method of R and

$$x_R = y + \lambda k x_R$$

in this case

$$x_R = (I - \lambda k)^{-1} y$$

where

$$x_R = (x_R(t_1), \ldots, x_R(t_n))^T$$

If we want to check the error, we will have:

$$e_n = x_R - x_n$$

In this case,

$$\|e_n\| \leq \frac{\|x_{n+1} - \boldsymbol{x}_n\|}{1 - \|\lambda k\|} \tag{5.8}$$

We also faced this error in the iterative method obtained from the Neumann series. Now for the integration rule R, the error will be as follows:

$$e_R(s) = x(s) - x_R(s)$$

And we have:

$$x_R(s) = y(s) + \lambda \int_a^b k(s,t) x_R(t) dt + E_R(\lambda k(s,t) x_R(t)) \tag{5.9}$$

We also have:

$$x(s) = y(s) + \lambda \int_a^b k(s,t) x(t) dt \tag{5.10}$$

From the difference of Eqs. (5.9) and (5.10), we will have:

$$e_R(s) = \lambda \int_a^b k(s,t)e_R(t)dt + E_R(\lambda k(s,t)x_R(t))$$

So, the function $e_R(s)$ holds in the second-type integral equation with the same solver kernel of $k(s,t)$ but with the second side function of $\lambda k(s,t)x(t)$, where s is a constant and t is a variable.

According to the above, for the integral equation in the form of $x = y + \lambda kx$, the answer is

$$x(s) = (I - \lambda k)^{-1} y(s) = (I + \lambda H)y(s)$$

So, according to rule R, for the above equation we can write:

$$\begin{aligned} e_R(s) &= (I - \lambda k)^{-1} E_R(\lambda k(s,t)x_R(t)) \\ &= (I + \lambda H)E_R(\lambda k(s,t)x_R(t)) \end{aligned}$$

Therefore

$$\|e_R\| \le (1 + \|\lambda H\|)\|E_R(\lambda k(s,t)x_R(t))\| \le \frac{\|E_R(\lambda k(s,t)x_R(t))\|}{1 - \|\lambda H\|} \tag{5.11}$$

where $\|\lambda k\| < 1$. According to Eq. (5.6), we have:

$$x(s) = y(s) + \lambda \sum_{j=1}^{n} w_j k(s,t_j)x(t_j) \tag{5.12}$$

and also

$$x_R(s) = y(s) + \lambda \sum_{j=1}^{n} w_j k(s,t_j)x_R(t_j) + E_R(\lambda k(s,t)x(t)) \tag{5.13}$$

From the difference of Eqs. (5.12) and (5.13), we can write:

$$e_R(s) = \lambda \sum_{j=1}^{n} w_j k(s,t_j)e_R(t_j) + E_R(\lambda k(s,t)x(t))$$

By calculating $e_R(t_i)$, we will have:

$$e_R(t_i) = \lambda \sum_{j=1}^{n} w_j k(t_i,t_j)e_R(t_j) + E_R(\lambda k(t_i,t)x(t)) \tag{5.14}$$

Assuming $k_i = E_R(\lambda k(t_i, t)x(t))$, Eq. (5.14) can be rewritten as follows

$$e_R = \lambda k e_R + k$$

By taking norms, we have

$$\|e_R\| = \|\lambda k\| \cdot \|e_R\| + \|k\|$$

as a result

$$\|e_R\| \le \frac{\|k\|}{1 - \|\lambda k\|}$$

Finally, the general error can be obtained as follows

$$\begin{aligned}\|x_n - x\| &= \|(x_n - x_R) + (x_R - x)\| \\ &\le \|x_n - x_R\| + \|x_R - x\|\end{aligned}$$

where $\|x_n - x_R\|$ and $\|x_R - x\| = \|e_R\|$ have been calculated in Eqs. (5.8) and (5.11).
Now, we will introduce the Nystrom method.

5.3 Nystrom Method

Suppose that

$$x(s) = y(s) + \lambda \int_a^b k(s, t)x(t)dt \tag{5.15}$$

And

$$x(s) = y(s) + \lambda \sum_{j=1}^{n} w_j k(s, t_j)x(t_j) \tag{5.16}$$

Assuming $s = t_i$ in Eq. (5.16), we have:

$$x(t_i) = y(t_i) + \lambda \sum_{j=1}^{n} w_j k(t_i, t_j)x(t_j)$$

So, we have the following matrix equation

$$x = y + \lambda kx$$

In comparison with the Neumann method, we conclude that the Neumann method also leads to the above equation by introducing iterative methods and writing integral equations corresponding to the integration methods. But it can be said that the Neumann method is in fact a generalized Jacobian method for solving the equation derived from the Nystrom method.

The iterative method can be written as follows:

$$(I - D(\lambda k))x_{n+1} = y + \lambda(k - D(k))x_n$$

By substituting $x_n = x_{n+1} = x$ in Eq. (5.18), we will have:

$$(I - D(\lambda k))x = y + \lambda(k - D(k))x$$

Therefore

$$x = y + \lambda kx \tag{5.17}$$

So, it can be said that

$$x_{n+1} = y + \lambda kx_n \tag{5.18}$$

is a iterative method for Eq. (5.17). But in general, Eq. (5.17) is solved by direct methods. Of course, sometimes the (5.18) methods are also faster. For one-dimensional problems, the problem (5.17) works well for an integration method of Gaussian type with 10–15 points, and it is not necessary to solve (5.18). However, if the problem is two-dimensional or more, or weak integration methods such as trapezoid rule are used, a large n should be adopted, and therefore it is necessary to use iterative methods of (5.18). In equation $(I - \lambda k)x = y$, a $n \times n$ system must be solved to calculate x. In this system, if λ is a regular value (even if it is almost regular, i.e., close to a characteristic value), then

$$x_1 = \lambda kx_1, x_1 is\, the, characteristic\, function$$

and if x_2 also satisfies $x = y(s) + \lambda kx$, it can be easily verified that $x_2 + \mu x_1$ also satisfies the above equation. So the solution is not unique. Thus, we can say that if λ is a regular value, every multiple of x_1 and x_2 is a solution to the equation. That is, $\left(I - \lambda \underline{k}\right)$ is singular for regular or near-regular λ.

We will now briefly review the error.

In this equation $(1 + \|\lambda k\|)^{-1}$ depends only on the kernel k and the value of λ, which are the inherent conditions of the problem. But $\|E_R(\lambda k(s, t)x_R(t))\|$ depends not only on the kernel k but also on the integration rule and the answer. If a proper integration rule is chosen, we expect this factor to be as small as desired.

By introducing the following theorem, we give a brief review of Gaussian numerical integration methods.

5.3.1 Theorem

If $k, x, y \in c[a, b]$, and $R_n$is a sequence of point wise integration rules such that

$$\lim_{n\to\infty} |If - R_n(f)| = 0$$

In this case, the Nystrom rule is also convergent.

The p-point composite rule, $NC(p, m)$, where p is the number of points and m is the number of subsets, has the above-mentioned property. The closed Gauss–Chebyshev method and the open Gauss–Chebyshev method in the form of

$$\int_{-1}^{1} \frac{f(x)}{\sqrt{1-x^2}} dx \simeq \frac{\pi}{n} \sum_{i=1}^{n} f(x_i)$$

$$= \frac{\pi}{n} \sum_{i=1}^{n} f\left(\cos \frac{(2i-1)\pi}{2n}\right)$$

also have this property.

5.4 Gauss–Chebyshev Method

To apply this method, we use Chebyshev polynomials.

The roots of these polynomials are used as integration points

$$T_n(x) = \cos\left(n\cos^{-1}x\right) = \cos n\theta$$

where $\theta = \cos^{-1}x$. To find the roots of $T_n(x) = 0$, we have:

$$\cos\left(n\cos^{-1}x\right) = 0$$

as a result:

$$x_i = \cos\theta_i, \theta_i = \frac{(2i-1)\pi}{2n}$$
$$x_i \neq \pm 1, \theta_i \neq 0, \pi, i = 1, \ldots, n$$

It can be shown that open Gauss–Chebyshev rule is accurate for polynomials of degree $n \neq 2kN, k \in \mathbb{N}$.

For the purpose, if we assume $f(x) = T_n(x)$ and $n \neq 2kN, k \in \mathbb{N}$, we have:

$$\int_{-1}^{1} \frac{T_n(x)}{\sqrt{1-x^2}} dx = \frac{\pi}{N} \sum_{i=1}^{N} \cos\frac{(2i-1)n\pi}{2N} \tag{5.19}$$

as a result

$$\int_{-1}^{1} \frac{T_i(x)T_j(x)}{\sqrt{1-x^2}} = \begin{cases} \pi, i = j = 0 \\ 0, i \neq j \\ \frac{\pi}{2}, i = j > 0 \end{cases}$$

Now if we assume $n = 0$, we will have:

$$\begin{aligned} \int_{-1}^{1} \frac{T_0(x)}{\sqrt{1-x^2}} dx &= \int_{-1}^{1} \frac{T_0(x) \times 1}{\sqrt{1-x^2}} dx \\ &= \int_{-1}^{1} \frac{T_0(x)T_0(x)}{\sqrt{1-x^2}} dx \\ &= \pi \end{aligned}$$

For $n > 0$, we have:

$$\begin{aligned} \int_{-1}^{1} \frac{T_n(x)}{\sqrt{1-x^2}} dx &= \int_{-1}^{1} \frac{T_n(x) \times 1}{\sqrt{1-x^2}} dx \\ &= \int_{-1}^{1} \frac{T_n(x)T_0(x)}{\sqrt{1-x^2}} dx \\ &= 0 \end{aligned}$$

and also this means that for $n \neq 2kN$, we have:

$$\frac{\pi}{N} \sum_{i=1}^{N} \cos\frac{(2i-1)n\pi}{2N} = 0, n = kN, k = 1, \ldots$$

Then, if $n \neq 2kN$, therefore

$$\sin\frac{n\pi}{2N} \neq 0$$

as a result

$$\begin{aligned}
2\sin\frac{n\pi}{2N}\sum_{i=1}^{N}\cos\frac{(2i-1)n\pi}{2N} &= \sum_{i=1}^{N} 2\sin\frac{n\pi}{2N}\cdot\cos\frac{(2i-1)n\pi}{2N} \\
&= \sum_{i=1}^{N}\left(\sin\frac{in\pi}{N} - \cos\frac{(i-1)n\pi}{N}\right) \\
&= \sin n\pi - \sin 0 = 0
\end{aligned}$$

thus

$$\sum_{i=1}^{N}\cos\frac{(2i-1)n\pi}{2N} = 0$$

5.4.1 Chebyshev Expansion

Suppose $f(x) = \sum_{i=0}^{\infty} a_i T_i(x)$, i.e., the function f is produced by orthogonal polynomials of $T_i(x)$, in which case it has a Chebyshev expansion and we have:

$$\int_{-1}^{1}\frac{f(x)}{\sqrt{1-x^2}}dx = \sum_{i=1}^{\infty} a_i \int_{-1}^{1}\frac{T_i(x)}{\sqrt{1-x^2}}dx$$

therefore

$$Ef = \int_{-1}^{1}\frac{f(x)}{\sqrt{1-x^2}}dx - \frac{\pi}{N}\sum_{i=1}^{N} f\left(\cos\frac{(2i-1)\pi}{2N}\right)$$

and we know that

$$ET_i = 0, i \neq 2kN$$

so

$$Ef = \sum_{i=0}^{\infty} a_i ET_i = \sum_{k=1}^{\infty} a_{2kN} ET_{2kN}$$

thus

$$ET_{2kN} = \int_{-1}^{1} \frac{T_{2kN}(x)}{\sqrt{1-x^2}} dx - \frac{\pi}{N} \sum_{i=1}^{N} \cos((2i-1)k\pi)$$

where

$$\sum_{i=1}^{N} \cos((2i-1)k\pi) = \begin{cases} -N, oddk \\ N, evenk \end{cases}$$

also

$$\int_{-1}^{1} \frac{T_{2kN}(x)}{\sqrt{1-x^2}} dx = 0, 2kN \neq 0$$

then

$$ET_{2kN} = \begin{cases} \pi, oddk \\ -\pi, evenk \end{cases}$$

therefore

$$Ef = \pi a_{2N} - \pi a_{4N} \pm \cdots$$

and as a result

$$|Ef| \leq \pi(|a_{2N}| + |a_{4N}| + \cdots)$$

If the convergence of Chebyshev expansion of f is fast, the dominant phrase in the error term will be πa_{2N}, i.e.,

$$|Ef| \simeq \pi |a_{2N}|$$

Now we want to obtain the expansion coefficients because without these coefficients, the function $f(x)$ cannot be known. For this purpose, according to the Chebyshev expansion, we can write:

$$f(x) = \sum_{i=0}^{\infty} a_i T_i(x)$$

therefore

$$\int_{-1}^{1} \frac{f(x)T_j(x)}{\sqrt{1-x^2}} dx = \sum_{i=0}^{\infty} a_i \int_{-1}^{1} \frac{T_i(x)T_j(x)}{\sqrt{1-x^2}} dx$$
$$= a_j \begin{cases} \frac{\pi}{2}, i = j \neq 0 \\ \pi, i = j = 0 \end{cases}$$

where

$$a_j = \begin{cases} \frac{2}{\pi} \int_{-1}^{1} \frac{f(x)T_j(x)}{\sqrt{1-x^2}} dx, \ j \neq 0 \\ \frac{1}{\pi} \int_{-1}^{1} \frac{f(x)T_j(x)}{\sqrt{1-x^2}} dx, \ j = 0 \end{cases}$$

If the above definite integrals can be calculated, the Gauss–Chebyshev numerical integration methods, i.e.,

$$a_j \simeq \frac{\pi}{N} \sum_{i=1}^{N} f\left(\cos\frac{(2i-1)\pi}{2N}\right) \cdot \cos\frac{(2i-1)\pi}{2N}$$

can be used to calculate a_{2N}.

5.4.2 Closed Gauss–Chebyshev Rule

The following rule is called the closed Gauss–Chebyshev rule:

$$\int_{-1}^{1} \frac{f(x)}{\sqrt{1-x^2}}dx = \frac{\pi}{N}\sum_{k=0}^{N}{}'' f\left(\cos\frac{k\pi}{N}\right)$$

The sign $''$ indicates the coefficient $\frac{1}{2}$ in the first and last terms of the summation. This rule is accurate for polynomials of at most degree $2N-1$. Therefore, we have:

$$E_c f = \int_{-1}^{1} \frac{f(x)}{\sqrt{1-x^2}}dx - \frac{\pi}{N}\sum_{k=0}^{N}{}'' f\left(\cos\frac{k\pi}{N}\right)$$

In this case, it can also be proved that

$$E_c T_i(x) = 0, i \neq 2jN, j \in \mathbb{N}$$

therefore

$$E_c f = \int_{-1}^{1} \frac{T_i(x)}{\sqrt{1-x^2}}dx - \frac{\pi}{N}\sum_{k=0}^{N}{}'' T_i\left(\cos\frac{k\pi}{N}\right)$$

For $i \neq 2jN$ we have:

$$E_c T_i(x) = 0 - \frac{\pi}{N}\sum_{k=0}^{N}{}'' \cos\frac{ik\pi}{N}$$

given that

$$2\sin\frac{i\pi}{N} \neq 0, i \neq jN$$

then

$$
\begin{aligned}
&2\sin\frac{i\pi}{N}\sum_{k=0}^{N}{}''\cos\frac{ik\pi}{N}=\sum_{k=0}^{N}{}''2\sin\frac{i\pi}{N}\cos\frac{ik\pi}{N}\\
&=\sin\frac{i\pi}{N}+(-1)^i\sin\frac{i\pi}{N}\\
&+\sum_{k=1}^{N-1}\left(\sin\frac{(k+1)i\pi}{N}-\sin\frac{(k-1)i\pi}{N}\right)\\
&=\sin\frac{i\pi}{N}+(-1)^i\sin\frac{i\pi}{N}\\
&+\sum_{k=1}^{N-1}\left(\sin\frac{(k+1)i\pi}{N}-\sin\frac{ki\pi}{N}\right)\\
&+\sum_{k=1}^{N-1}\left(\sin\frac{ik\pi}{N}-\sin\frac{(k-1)i\pi}{N}\right)\\
&=\sin\frac{i\pi}{N}+(-1)^i\sin\frac{i\pi}{N}+\sin i\pi\\
&-\sin\frac{i\pi}{N}+\sin\frac{(N-1)i\pi}{N}+\sin 0\\
&=(-1)^i\sin\frac{i\pi}{N}+\sin\left(i\pi-\frac{i\pi}{N}\right)\\
&=\begin{cases}-\sin\frac{i\pi}{N}+\sin\frac{i\pi}{N}=0, oddi\\ \sin\frac{i\pi}{N}-\sin\frac{i\pi}{N}=0, eveni\end{cases}
\end{aligned}
$$

so

$$\sum_{k=0}^{N}{}''\cos\frac{ik\pi}{N}=0$$

then

$$E_cT_i(x)=0$$

i.e.,

$$\int_{-1}^{1}\frac{T_i(x)}{\sqrt{1-x^2}}dx=\frac{\pi}{N}\sum_{k=0}^{N}{}''\cos\frac{ik\pi}{N}$$

In the case that $i=2jN$

$$
\begin{aligned}
E_c T_{2jN}(x) &= 0 - \frac{\pi}{N} \sum_{k=0}^{N}{}'' \cos \frac{ik\pi}{N} \\
&= -\frac{\pi}{N} \sum_{k=0}^{N}{}'' 1 \\
&= -\pi
\end{aligned}
$$

by assuming

$$
f(x) = \sum_{i=0}^{\infty}{}'' b_i T_i(x)
$$

we have

$$
\begin{aligned}
E_c f &= \int_{-1}^{1} \frac{f(x)}{\sqrt{1-x^2}} dx - \frac{\pi}{N} \sum_{k=0}^{N}{}'' f\left(\cos \frac{k\pi}{N}\right) \\
&= \sum_{i=0}^{\infty}{}'' b_i E(T_i(x)) \\
&= \sum_{i=0}^{\infty}{}'' b_i \left(\int_{-1}^{1} \frac{T_i(x)}{\sqrt{1-x^2}} dx - \frac{\pi}{N} \sum_{k=0}^{N}{}'' f\left(\cos \frac{ik\pi}{N}\right) \right) \\
&= \sum_{j=0}^{\infty}{}'' - \pi b_{2jN}
\end{aligned}
$$

therefore

$$
|E_c f| \leq x \sum_{j=1}^{\infty} \left| b_{2jN} \right| \simeq \pi |b_{2N}|
$$

It can be concluded that.

(1) Both formulas are accurate for polynomial functions of at most $2N - 1°$.
(2) The open formula uses the value of the function at N points and the closed formula uses $N + 1$ points.
(3) If $f(\pm 1) = 0$, the closed formula uses $N + 1$ points.
(4) Closed formula's points are easier to calculate.

5.4.3 Theorem

If $f \in C^l[-1,1]$ *and* $f^{(l+1)}(x)$ *are continuous on* $[-1,1]$ *unless at a finite number of points, then there exists a* c_f *such that*

$$|b_i| \leq c_f i^{-l-1} = c_f i^{-p}$$

Now, using the Theorem (5.3.3), we have:

$$|E_c f| \leq \pi \sum_{j=1}^{\infty} |b_{2jN}| \leq \pi c_f \sum_{j=1}^{\infty} (2jN)^{-p},\ p > 1 \tag{5.20}$$

because

$$\begin{aligned}
\sum_{j=1}^{\infty}(2jN)^{-p} &= (2N)^{-p} + (4N)^{-p} + \cdots \\
&= (2N)^{-p}(1 + 2^{-p} + 3^{-p} + \cdots) \\
&= (2N)^{-p}\left(1 + \sum_{j=2}^{\infty} j^{-p}\right)
\end{aligned}$$

on the other hand

$$\sum_{j=2}^{\infty} j^{-p} < \int_1^{\infty} \frac{dx}{x^p},\ p > 1$$

where

$$\int_1^{\infty} \frac{dx}{x^p} = \int_1^{\infty} x^{-p} dx = \frac{1}{p-1}$$

then

$$\sum_{j=2}^{\infty} j^{-p} < \frac{1}{p-1}$$

therefore

$$\sum_{j=1}^{\infty}(2jN)^{-p} \le (2N)^{-p}\left(1+\frac{1}{p-1}\right)$$
$$= \frac{p}{p-1}(2N)^{-p} \tag{5.21}$$

As a result, according to Eqs. (5.20) and (5.21), we have:

$$|E_c f| \le \frac{p \cdot \pi \cdot c_f}{p-1}(2N)^{-p} = C \cdot N^{-p}$$

where

$$C = \frac{2^{-p} \cdot p \cdot c_f \cdot \pi}{p-1}$$

To calculate the upper bound parameters, we generally perform the following steps.

Suppose a sequence such as $\{a_n\}_{n=1}^{N}$ is present so that $\{|a_n|\}$ is descending and $|a_n| \le C \cdot n^{-r}$. We obtain C and r. For given $a_1, a_2, \ldots, a_N$, we form the following system:

$$\begin{cases} |a_n| = C \cdot n^{-r}, n = 1, \ldots, N \\ \log|a_n| = \log C - r\log n, n = 1, \ldots, N \\ \log C = \alpha, \log n = d_n, \log|a_n| = C_n \end{cases}$$

where

$$C_n = \alpha - r d_n, n = 1, \ldots, N$$

We calculate the least squares solution of this system. Generally, we multiply the obtained α in $\lambda, 2 \le \lambda \le 4$ and put it equal to $\log C$.

5.4.4 Disadvantages of the Gauss–Chebyshev Method

If the weight function does not exist, in order to use the Gauss–Chebyshev method, we must create the weight function. That means

$$\int_{-1}^{1} f(x)dx = \int_{-1}^{1} \frac{f(x)\sqrt{1-x^2}}{\sqrt{1-x^2}}dx = \int_{-1}^{1} \frac{\overline{f}(x)}{\sqrt{1-x^2}}dx$$

On the other hand, we know that the errors of the Chebyshev's open and closed methods depend on the continuous derivatives of $\overline{f}(x) = f(x)\sqrt{1-x^2}$. This function and its first derivative are not continuous on ± 1, so the condition of the theorem does not hold. Therefore, in the absence of a weight function, these integration methods are not accurate and cannot be computed for large N. In such cases, the Clenshaw–Curtis integration method is used. Suppose $f(x) = \sum_{i=0}^{\infty}{}' b_i T_i(x)$, therefore:

$$\int_{-1}^{1} f(x)dx = \sum_{i=0}^{\infty}{}' b_i \int_{-1}^{1} T_i(x)$$
$$= \sum_{i=0}^{\infty}{}' b_i \times \begin{cases} \int_{-1}^{1} dx = 2, i = 0 \\ \int_{-1}^{1} x dx = 0, i = 1 \end{cases}$$

Given that $T_i(x) = \cos i\theta$ and $x = \cos\theta$, $i > 1$ we have

$$\int_{-1}^{1} T_i(x)dx = \int_{0}^{\pi} \cos i\theta \cdot \sin\theta d\theta$$
$$= \frac{1}{2}\int_{0}^{\pi} (\sin(i+1)\theta - \sin(i-1)\theta)d\theta$$
$$= \frac{1}{2}\left[-\frac{\cos(i+1)\theta}{i+1} + \frac{\cos(i-1)\theta}{i-1}\right]_0^{\pi}$$
$$= \frac{1}{2}\left(-\frac{(-1)^{i+1}}{i+1} + \frac{(-1)^{i-1}}{i-1} + \frac{1}{i+1} - \frac{1}{i-1}\right)$$
$$= \begin{cases} 0, odd i \\ \frac{2}{1-i^2}, even i \end{cases}$$

So it can be said that

$$\int_{-1}^{1} f(x)dx = \sum_{i=0, i=2k}^{\infty}{}' \frac{2b_i}{1-i^2}, k \in \mathbb{N}$$

where as mentioned earlier, we have:

$$b_i = \frac{2}{\pi}\int_{-1}^{1} \frac{f(x)T_i(x)}{\sqrt{1-x^2}} dx$$

If the above integral cannot be analytically computed, the closed Gauss–Chebyshev method can be used to approximate it.

$$\int_{-1}^{1} \frac{f(x)T_i(x)}{\sqrt{1-x^2}} \simeq \frac{\pi}{N}\sum_{k=0}^{N}{}'' f\left(\cos\frac{k\pi}{N}\right)T_i\left(\cos\frac{k\pi}{N}\right)$$
$$= \frac{\pi}{N}\sum_{k=0}^{N}{}'' f\left(\cos\frac{k\pi}{N}\right)\cos\frac{ki\pi}{N}$$

which assuming

$$b_i \simeq \frac{2}{N}\sum_{k=0}^{N}{}'' f\left(\cos\frac{k\pi}{N}\right)\cos\frac{ik\pi}{N}$$

we will have

$$\int_{-1}^{1} f(x)dx \simeq \frac{4}{N}\sum_{i=0}^{\infty}{}' \frac{1}{1-i^2}\sum_{k=0}^{N}{}'' f\left(\cos\frac{k\pi}{N}\right)\cos\frac{ik\pi}{N}$$
$$\simeq \frac{4}{N}\sum_{i=0}^{N}{}'' \frac{1}{1-i^2}\sum_{k=0}^{N}{}'' f\left(\cos\frac{k\pi}{N}\right)\cos\frac{ik\pi}{N}$$
$$\simeq \sum_{k=0}^{N}{}'' \left(\frac{4}{N}\sum_{i=0}^{N}{}'' \cos\frac{ik\pi}{N}\cdot\frac{1}{1-i^2} f\left(\cos\frac{k\pi}{N}\right)\right)$$
$$= \sum_{k=0}^{N}{}'' w_k f\left(\cos\frac{k\pi}{N}\right)$$

where

$$w_k = \frac{4}{N}\sum_{i=0}^{N}{}'' \cos\frac{ik\pi}{N}\cdot\frac{1}{1-i^2}$$

5.5 Non-singular Functions

Numerical methods such as the Trapezoid or Simpson quadrature rules cannot be used to calculate the following integrals.

$$\int_0^2 \frac{dx}{\sqrt{x}}, \int_{-1}^1 \frac{dx}{\sqrt{1-x^2}}$$

and also

$$\int_0^1 \sqrt{x}dx$$

because they face no problem in the integration interval, but their derivatives face a problem in zero, so the error cannot be bounded to a small numerator. Consider the following integral.

$$\int_a^b w(x)f(x)dx$$

where $w(x)$ is a singular weight function and $f(x)$ is relatively smooth. The following situations can occur:

(1) The function has fast variations (this happens less often), like the function

$$\int_a^b w(x)f(x)dx$$

(2) Function f has a discontinuity in the finite number of points.
(3) Function f is singular.

For second case, suppose that $x_0, \ldots, x_n$ are singular points, therefore

$$\int_{x_0}^{x_1} f(x)dx + \int_{x_1}^{x_2} f(x)dx + \cdots = \sum_{i=0}^{n-1} \int_{x_i}^{x_{i+1}} f(x)dx$$

In calculating each of the above integrals, we employ a quadrature rule that does not use the initial and end of the interval.

For the third case, we have two methods:

(1) There is a Gaussian integration rule whose weight function is $w(x)$, like the Gauss–Chebyshev rule with $w(x) = \frac{1}{\sqrt{1-x^2}}$.
(2) We create a Chebyshev-type rule with the weight function $w(x)$ and use it. Suppose that the function $f(s)$ has the McLaren expansion of $f(x) = \sum_{i=0}^{\infty} a_i x^i$, therefore

$$\int_a^b w(x)f(x)dx = \sum_{i=0}^{\infty} a_i \int_a^b w(x)x^i dx$$
$$= \sum_{i=0}^{\infty} a_i m_i$$

where

$$m_i = \int_a^b w(x)x^i dx, i = 0, 1, \ldots$$

By analytically calculating m_i, the value of $\int_a^b w(x)f(x)dx$ can be calculated.

5.5.1 Definition

The sequence of functions $\{f_i(x)\}_{i=1}^{\infty}$ *is called linearly independent whenever*

$$\forall n \forall x(\alpha_1 f_1(x) + \cdots + \alpha_n f_n(x) = 0 \Rightarrow \alpha_1 = \cdots = \alpha_n = 0)$$

5.5.2 Definition

The sequence of functions $\{h_i(x)\}_{i=1}^{\infty}$ *is said complete whenever*

$$\forall f \forall i(\langle f(x), h_i(x)\rangle = 0 \Rightarrow f(x) = 0)$$

In general, if we write the expansion of $f(x)$ in terms of complete sequence functions, i.e.,

$$f(x) = \sum_{i=0}^{\infty} a_i h_i(x)$$

where $\{h_i(x)\}_i$ form a complete sequence of functions. In this case

$$\int_a^b w(x) f(x)dx = \sum_{i=0}^{\infty} a_i \int_a^b w(x)h_i(x)dx$$

$$= \sum_{i=0}^{\infty} a_i m_i$$

where

$$m_i = \int_a^b w(x)h_i(x)dx, i = 0, 1, \ldots$$

5.5.3 Example

Consider the sequence of $\{T_i(x)\}_i$. Suppose

$$f(x) = \sum_{i=0}^{\infty}{}' a_i T_i(x)$$

where

$$a_i = \frac{2}{\pi} \int_{-1}^{1} \frac{f(x)T_i(x)}{\sqrt{1-x^2}} dx$$

If we use the $N+1$-point rule to calculate a_i, we will have:

$$a_i \simeq \frac{2}{\pi} \sum_{k=0}^{N}{}'' f\left(\cos \tfrac{k\pi}{N}\right) \cos \frac{ik\pi}{N} = \overline{a}_i$$

thus

$$\int_a^b w(x) f(x)dx = \sum_{i=0}^{\infty}{}' a_i \int_a^b w(x)T_i(x)$$

$$= \sum_{i=0}^{\infty} a_i m_i$$

$$\simeq \sum_{i=0}^{N}{}' \overline{a}_i m_i$$

If $f(x)$ is a well-behaved function and $w(x)$ is a bad-behaved function, the singular point method is one way to escape being singular. Suppose we face a problem at point x_0, therefore

$$\int_a^b w(x)f(x)dx = \int_a^b w(x)(f(x) - f(x_0))dx + \int_a^b w(x)f(x_0)dx$$

$$= \int_a^b w(x)g(x)dx + f(x_0)\int_a^b w(x)dx$$

In this case, usually $g(x_0) = 0$ and $w(x_0)$ also face problems due to the behavior of $w(x)$.

5.5.4 Example

$$\int_0^1 s^{-\frac{1}{2}} f(x)ds = \int_0^1 s^{-\frac{1}{2}}(f(s) - f(0))ds + f(0)\int_0^1 s^{-\frac{1}{2}}ds$$

where

$$f(s) - f(0) \to 0, s \to 0$$

This convergence is usually a linear convergence, i.e., in the vicinity of zero

$$f(s) - f(0) \simeq ks$$

This does not remove the singularity, but weakens it, and new functions may have singular derivatives.

5.5.5 Example

The sequence $\{\sin ix\}_{i=1}^{\infty}$ is linearly independent on the interval $[0, 2\pi]$ but it is not complete. Because if $f(x) = 1$, then

$$\int_a^b f(x) f_i(x) dx = \int_0^{2\pi} 1 \times \sin i\, x dx = 0$$

While

$$f(x) \neq 0$$

5.5.6 *Example*

The sequence $\{\cos i\, x\}_{i=1}^{\infty}$ is linearly independent on the interval $[0, 2\pi]$ but it is not complete. Because if $f(x) = 1$, then

$$\int_a^b f(x) f_i(x) dx = \int_0^{2\pi} 1 \times \cos i\, x dx = 0$$

While

$$f(x) \neq 0$$

It can also be shown by using the contraposition of the Definition (5.4.2) that the set

$$\{1, \cos x, \cos 2x, \ldots, \sin x, \sin 2x, \ldots\}$$

is not a complete set.

Note: If a sequence contains a complete subsequence, it is complete itself.

We now use the above discussion as an expansion method for solving integral equations.

5.6 Expansion Method

Suppose

$$g(s) = x(s) + \lambda \int_a^b k(s, t) x(t) dt \tag{5.22}$$

Also assume that sequence $\{h_i(x)\}_{i=1}^{\infty}$ is linearly independent and complete. Using the expansion method of

$$x(s) \simeq x_n(s) = \sum_{i=1}^{n} a_i h_i(s) \forall_n \tag{5.23}$$

and substituting the Eq. (5.23) in the Eq. (5.22), we have:

$$g(s) \simeq \sum_{j=1}^{n} a_j h_j(s) + \lambda \sum_{j=1}^{n} a_j \int_a^b k(s,t) h_j(t) dt$$

So

$$g(s) \simeq \sum_{j=1}^{n} a_j \left(h_j(s) + \lambda \int_a^b k(s,t) h_j(t) dt \right) \tag{5.24}$$

Equation (5.24) is a linear system. It is enough to calculate a_j s as a vector.

5.7 Collocation Methods

One of the methods that we can employ is the collocation method, which means the method of using points. We obtain the system (5.21) for s_i, $i = 1, \ldots, n$. We will have

$$g(s_i) \simeq \sum_{j=1}^{n} a_j \left(h_j(s_i) + \lambda \int_a^b k(s_i,t) h_j(t) dt \right)$$

Assuming

$$c_{ij} = h_j(s_i) + \lambda \int_a^b k(s_i,t) h_j(t) dt, i, j = 1, \ldots, n \tag{5.25}$$

we have

$$g_i = \sum_{j=1}^{n} c_{ij} a_j, i = 1, \ldots, n \tag{5.26}$$

The matrix form of Eq. (5.23) will be as follows

$$Ca = g \tag{5.27}$$

where

$$C = \left[c_{ij}\right]_{i,j=1}^{n}, g = [g_1, \ldots, g_n], a = [a_1, \ldots, a_n]$$

if

$$\int_a^b k(s_i, t)h_j(t)dt, 1 \leq i, j \leq n$$

can be calculated analytically, then c_{ij} s will be calculated accurately and the system of Eq. (5.24) will have an answer as a.

It should be noted that if λ is a regular value, the matrix C must be non-singular. Also, it is possible that λ is not a regular value, but the matrix C still is non-singular. It is possible that if C is not close to being non-singular, λ is not a regular value, and this is due to the converting approximation to equality.

5.7.1 Example

Suppose that

$$10s + 6 = x(s) + 18\int_0^1 (s + t)x(t)dt \tag{5.28}$$

Assuming $h_i(s) = s^{i-1}$, we have

$$x(s) \simeq x_3(s) = a_1 + a_2 s + a_3 s^2$$

Also according to Eq. (5.22), we have

$$c_{ij} = s_i^{j-1} + 18\int_0^1 (s_i + t)t^{j-1}dt, i, j = 1, 2, 3$$

Consider points $s_1 = 0, s_2 = \frac{1}{2}$ and $s_3 = 1$. By calculating c_{ij} s, the system (5.27) is obtained as follows

$$\begin{bmatrix} 10 & 7 & \frac{11}{2} \\ 19 & 11 & \frac{31}{4} \\ 28 & 16 & \frac{23}{2} \end{bmatrix} \begin{bmatrix} a_1 \\ a_2 \\ a_3 \end{bmatrix} = \begin{bmatrix} 6 \\ 11 \\ 16 \end{bmatrix}$$

By solving the system, we have

$$a_1 = a_3 = 0, a_2 = 1$$

And as a result, $x(s) = s$ is obtained which is the exact answer.

Note: If we guess the oddness or evenness of the integral equation answer at the first, we can choose $x(s)$ with respect to the odd or even functions. For example, in the above example

$$x_1(s) = a_2 s^1 = a_2 h_2(s), c_{21} = s_1 + 18 \int_0^2 (s_1 + t) t dt = 10 s_1 + 6$$

In this case, the system (5.29) is as follows

$$c_{21} a_2 = g(s_1)$$

As a result, $a_2 = 1$.

For the second solution, the integral in the Eq. (5.25) cannot be easily computed, so we have to approximate it.

$$\int_a^b k(s,t) h_j(t) dt \simeq \sum_{k=1}^{m} w_{ik} k(s, t_i) h_j(t_i), i = 1, \ldots, m \tag{5.29}$$

If we substitute the Eq. (5.29) in Eq. (5.24), we will have

$$g(s) \simeq \sum_{j=1}^{n} a_j \left(h_j(s) + \lambda \sum_{k=1}^{m} w_{ik} k(s, t_i) h_j(t_i) \right), i = 1, \ldots, m$$

For $s = t_u$, $u = 1, \ldots, n$ and $m = n$ we have

$$g(t_u) \simeq \sum_{j=1}^{n} a_j \left(h_j(t_u) + \lambda \sum_{k=1}^{n} w_{ik} k(t_u, t_i) h_j(t_i) \right), i = 1, \ldots, n$$

assuming

$$b_{ij} = h_j(t_u) + \lambda \sum_{k=1}^{n} w_{ik} k(t_u, t_i) h_j(t_i)$$

and

$$B = [b_{ij}]_{i,j=1}^{m}$$

The system (5.25) changes to the following system:

$$Ba = g$$

In the Example (5.6.1), we apply the Simpson rule with $g = \frac{1}{2}$ to calculate the integral of Eq. (5.28). Then

$$x(s) \simeq x_2(s) = a_1 + a_2 s$$

$$\int_0^1 (s+t)dt = \frac{1}{6}\left(s + 4\left(s + \frac{1}{2}\right) + (s+1)\right)$$

Using the above equations and assumption of $s = 0, 1$, we have the following system

$$\begin{cases} 10a_1 + 6a_2 = 6 \\ 28a_1 + 16a_2 = 16 \end{cases}$$

Finally, $a_1 = 0$ and $a_2 = 1$. Since the Simpson rule for functions of at most degree three is accurate, the answer is accurate. In general, it can be written

$$x = y + \lambda k x$$

where assuming $\lambda = 1$ we have,

$$x = y + kx$$

as a result

$$(I - k)x = y$$

Assuming $L = I - k$ we have

$$Lx = y$$

To calculate the error and estimate some of its limits, assume that if $x \simeq x_n$, then

$$r_n = Lx_n - y = Lx_n - y + (y - Lx)$$

therefore

$$r_n = L(x_n - x)$$

and if $\varepsilon_n = x_n - x$ we have

$$r_n = L\varepsilon_n$$

therefore

$$\|r_n\| \leq \|L\| \cdot \|\varepsilon_n\|, \|L\| = \|I - k\| \leq 1 + \|k\|$$

then

$$\|r_n\| \leq (1 + \|k\|)\|\varepsilon_n\|$$

as a result

$$\|\varepsilon_n\| \geq \frac{\|r_n\|}{1 + \|k\|}$$

given that

$$r_n = (I - k)\varepsilon_n = \varepsilon_n - k\varepsilon_n$$

so

$$\varepsilon_n = k\varepsilon_n + r_n$$

then

$$\|\varepsilon_n\| \leq \|k\| \cdot \|\varepsilon_n\| + \|r_n\|$$

Assuming $\|k\| < 1$ we have

$$(1 - \|k\|)\|\varepsilon_n\| \leq \|r_n\|$$

therefore

$$\frac{\|r_n\|}{1+\|k\|} \leq \|\varepsilon_n\| \leq \frac{\|r_n\|}{1-\|k\|}$$

It can be seen that if $\|k\|$ is close to 1, $\frac{\|r_n\|}{1-\|k\|}$ will be large, although if $\|r_n\|$ is small, it is not a reason for $\|\varepsilon_n\|$ to be small. But if $\|r_n\| \to 0$, then, $\|\varepsilon_n\| \to 0$.

now if we have

$$x(s) = y(s) + \int_a^b k(s,t)x_n(t)dt, \; x(s) \simeq x_n(s)$$

therefore

$$r_n(s) = y(s) - x_n(s) + \int_a^b k(s,t)x_n(t)dt$$

then

$$\begin{aligned} r_n(s) &= y(s) - \sum_{i=1}^{n} a_i^{(n)} h_i(s) + \sum_{i=1}^{n} a_i^{(n)} \int_a^b k(s,t)h_i(t)dt \\ &= y(s) - \sum_{i=1}^{n} a_i^{(n)} \left(h_i(s) - \int_a^b k(s,t)h_i(t)dt \right) \end{aligned}$$

if

$$k_i(s) = \int_a^b k(s,t)h_i(t)dt$$

then

$$h_i(s) - k_i(s) = l_i(s) \tag{5.30}$$

So

$$r_n = y(s) - \sum_{i=1}^{n} a_i^{(n)} l_i(s)$$

where the goal is to estimate $a_1^{(n)}, \ldots, a_n^{(n)}$.

5.8 Norm Chebyshev

One method is $\min_{\underline{a}^{(n)}} \|r_n\|$ where the applied norm is Chebyshev norm.

Then

$$\|x\| = \max_{a \le s \le b} \|x(s)\|$$

And

$$\|k\| = \max_{a \le s \le b} \int_a^b |k(s, t)| dt$$

To calculate $\|x\|$, it is not always possible to calculate the maximum, for example, it is possible that the function is not differentiable. In this case, we consider a series of points (e.g., equidistant points) and calculate the value of the function at those points and obtain the maximum absolute value. That is, we assume

$$A = \{s_i | i = 1, \ldots, q\}$$

To calculate $\|k\|$, one method for calculating the integral is the Monte Carlo method. Since $0 < RND < 1$, therefore $0 < RND(b - a) < b - a$, and as a result

$$a < RND(b - a) + a < b$$

where with these points

$$\int_a^b f(x) dx \simeq \frac{b - a}{n} \sum_{i=1}^{n} y(s_i)$$

So, for calculating $\min_{\underline{a}^{(n)}} \|r_n\|$, it is enough to calculate

$$\min_{\underline{a}^{(n)}} \max_{s_i \in A} \left| y(s_i) - \sum_{j=1}^{n} a_j^{(n)} l_j(s_i) \right|$$

5.9 Least Squares Method (L_2-Norm Method)

Assuming that this method is familiar to readers, without a basic explanation for the method, we will examine it directly in the integral equation. This method is for real functions and

$$\min_{\underline{a}^{(n)}} \|r_n\| \int_a^b \left(y(s) - \sum_{i=1}^{n} a_i^{(n)} l_i(s) \right)^2 ds$$

By assuming

$$I\left(\underline{a}^{(n)}\right) = \int_a^b \left(y(s) - \sum_{i=1}^{n} a_i^{(n)} l_i(s) \right)^2 ds \tag{5.31}$$

It can be written

$$\min_{\underline{a}^{(n)}} \left(\underline{a}^{(n)}\right)$$

For this purpose, we put

$$\frac{\partial I\left(\underline{a}^{(n)}\right)}{\partial_{a_i}} = 0$$

According to Eq. (5.31), we have

$$I\left(\overset{(n)}{a}\right) = \int_a^b \left(y^2(s) - 2y(s) \sum_{i=1}^{n} a_i^{(n)} l_i(s) \right) ds$$
$$+ \int_a^b \left(\sum_{i=1}^{n} a_i^{(n)} l_i(s) \times \sum_{j=1}^{n} a_j^{(n)} l_j(s) \right) ds$$

therefore

$$\frac{\partial I\left(\overset{(n)}{a}\right)}{\partial a_i} = \int_a^b 2 \sum_{j=1}^{n} a_j^{(n)} l_j(s) l_i(s) ds - 2 \int_a^b y(s) l_i(s) ds$$
$$= 0, i = 1, \ldots, n$$

we define

$$(Y_{ls})_i = \int_a^b y(s)l_i(s)ds, \tag{5.32}$$

$$(L_{ls})_{ij} = \int_a^b l_i(s)l_j(s)ds$$

so we have

$$\sum_{j=1}^{n} (L_{ls})_{ij} \cdot a_j^{(n)} = (Y_{ls})_i, i = 1, \ldots, n \tag{5.33}$$

And the matrix form of the Eq. (5.33) is as follows

$$L_{ls}\underline{a}^{(n)} = Y_{ls}$$

where

$$l_i(s) = h_i(s) - \int_a^b k(s,t)h_i(t)dt$$

And calculating such integrals can be very difficult.

5.9.1 Example

Consider the following equation

$$x(s) = \sin s - \frac{1}{4}s + \frac{1}{4}\int_0^{\frac{\pi}{2}} stx(t)dt$$

with the exact answer of $x(s) = \sin s$. Given that $x(s)$ is an odd function, suppose

$$h_1(s) = s, h_2(s) = s^3$$

Then

$$x(s) \simeq x_2(s) = a_1 s + a_2 s^3$$

And according to Eq. (5.32), we have

$$l_1(s) = h_1(s) - \frac{1}{4}\int_0^{\frac{\pi}{2}} sth_1(t)dt$$

$$= s - \frac{1}{4}\int_0^{\frac{\pi}{2}} st^2(t)dt$$

$$= s\left(1 - \frac{\pi^3}{96}\right)$$

$$l_2(s) = h_2(s) - \frac{1}{4}\int_0^{\frac{\pi}{2}} sth_2(t)dt$$

$$= s^3 - \frac{1}{4}\int_0^{\frac{\pi}{2}} st^4(t)dt$$

$$= s\left(s^2 - \frac{\pi^5}{640}\right)$$

And according to Eq. (5.32), we have

$$(L_{ls})_{11} = \int_0^{\frac{\pi}{2}} s^2\left(1 - \frac{\pi^3}{96}\right)^2 ds$$

$$\simeq 0.59215955$$

$$(L_{ls})_{21} = (L_{ls})_{12} = \int_0^{\frac{\pi}{2}} s^2\left(1 - \frac{\pi^3}{96}\right) \cdot s\left(s^2 - \frac{\pi^5}{640}\right) ds$$

$$\simeq 0.87665709$$

$$(L_{ls})_{22} = \int_0^{\frac{\pi}{2}} s\left(s^2 - \frac{\pi^5}{640}\right) ds$$

$$\simeq 1.83717689$$

And

$$(Y_{ls})_1 = \int_0^{\frac{\pi}{2}} s\left(1 - \frac{\pi^3}{96}\right)\left(\sin s - \frac{1}{4}s\right) ds$$

$$\simeq 0.45835530$$

$$(Y_{ls})_2 = \int_0^{\frac{\pi}{2}} s\left(s^2 - \frac{\pi^5}{640}\right)\left(\sin s - \frac{1}{4}s\right)ds$$

$$\simeq 0.6002751$$

So, we form the following system

$$\begin{bmatrix} 0.59215955 & 0.87665709 \\ 0.87665709 & 1.83717689 \end{bmatrix}\begin{bmatrix} a_1 \\ a_2 \end{bmatrix} = \begin{bmatrix} 0.45835530 \\ 0.60032751 \end{bmatrix}$$

By solving this system, we will have

$$a_1 \simeq 0.9887922, a_2 \simeq -0.1450618$$

where according to

$$x(s) = \sin s = s - \frac{s^3}{3!} + \frac{s^5}{5!} - \cdots$$

We have

$$a_1 = 1, a_2 = -\frac{1}{6} = -0.16666\ldots$$

5.10 Numerical Solution of the Second Kind Integral Equations

Consider the following integral equation

$$x(s) = y(s) + \int_a^b k(s,t)x(t)dt, a \leq s \leq b \tag{5.34}$$

If we solve the integral existed in the above equation using the corrected composite trapezoidal quadrature method, we will have

$$x(s) = y(s) + \frac{h}{2}k(s, s_0) + h\sum_{j=1}^{n-1} k(s, s_j)x(s_j)$$

$$\frac{h}{2}k(s, s_n)x(s_n) + \frac{h^2}{12}\big(J(s, s_0)x(s_0) + k(s, s_0)x'(s_0)\big)$$
$$- \frac{h^2}{12}\big(J(s, s_n)x(s_n) + k(s, s_n)x'(s_n)\big) \tag{5.35}$$

where

$$J(s, t)|_{t=s_j} = \left[\frac{\partial k(s, t)}{\partial t}\right]_{t=s_j}, j = 0, 1, \ldots, n$$

Now if we substitute $s = s_i$ in Eq. (5.35), we will have

$$x(s_i) = y(s_i) + \frac{h}{2}k(s_i, s_0)x(s_0) + h\sum_{j=1}^{n-1} k(s_i, s_j)x(s_j)$$
$$+ \frac{h}{2}k(s_i, s_n)x(s_n) + \frac{h^2}{12}\big(J(s_i, s_0)x(s_0) + k(s_i, s_0)x'(s_0)\big)$$
$$- \frac{h^2}{12}\big(J(s_i, s_n)x(s_n) + k(s_i, s_n)x'(s_n)\big), i = 0, \ldots, n \tag{5.36}$$

If we take the derivative of Eq. (5.34) with respect to s, we have

$$x'(s) = y'(s) + \int_a^b m(s, t)x(t)dt \tag{5.37}$$

where

$$m(s, t) = \frac{\partial k(s, t)}{\partial s}$$

The answer function of $x(s)$, which satisfies the Eq. (5.34), also holds for the Eq. (5.36). Consider Eq. (5.37). The following situations can occur:

(1) $\frac{\partial^2 k(s,t)}{\partial s \partial t}$ is not available
(2) $\frac{\partial^2 k(s,t)}{\partial s \partial t} = l(s, t)$ are available.

In the case (1), we solve the Eq. (5.36) using the composite trapezoidal quadrature method. In this case, we will have

$$x'(s) = y'(s) + \frac{h}{2}m(s, s_0)x(s_0) + h\sum_{j=1}^{n-1} m(s, s_j)x(s_j) + \frac{h}{2}m(s, s_n)x(s_n) \tag{5.38}$$

Now by substituting $s = s_i$ in Eq. (5.40), we will have

$$x'(s_i) = y'(s_i) + \frac{h}{2}m(s_i, s_0)x(s_0) + h\sum_{j=1}^{n-1} m\big(s_i, s_j\big)x\big(s_j\big)$$
$$+\frac{h}{2}m(s_i, s_n)x(s_n),\ i = 0, \ldots, n \tag{5.39}$$

From the system (5.36), we have

$$x(s_i) = y(s_i) + \left(\frac{h}{2}k(s_i, s_0) + \frac{h^2}{12}J(s_i, s_0)\right)x(s_0)$$
$$+h\sum_{j=1}^{n-1} k(s_i, s_0)x\big(s_j\big) + \left(\frac{h}{2}k(s_i, s_n) - \frac{h^2}{12}J(s_i, s_n)\right)x(s_n)$$
$$+\frac{h^2}{12}\big(k(s_i, s_0)x'(s_0) - k(s_i, s_n)x'(s_n)\big),\ i = 0, \ldots, n \tag{5.40}$$

By setting $i = 0$ and $i = n$ in Eq. (5.36), we also have:

$$x'(s_0) = y'(s_0) + \frac{h}{2}m(s_0, s_0)x(s_0)$$
$$+h\sum_{j=1}^{n-1} m\big(s_0, s_j\big)x\big(s_j\big) + \frac{h}{2}m(s_0, s_n)x(s_n)$$
$$x'(s_n) = y'(s_n) + \frac{h}{2}m(s_n, s_0)x(s_0)$$
$$+h\sum_{j=1}^{n-1} m\big(s_n, s_j\big)x\big(s_j\big) + \frac{h}{2}m(s_n, s_n)x(s_n) \tag{5.41}$$

From Eqs. (5.40) and (5.41), a system of $n + 3$ equations with $n + 3$ unknowns is obtained and the answer of this system is $x(s_0), \ldots, x(s_n), x'(s_0), x'(s_n)$, which is the approximate answer calculated in points $s_0, \ldots, s_n$ for Eq. (5.34).

In case (2), we rewrite Eq. (5.37) using the corrected trapezoidal rule.

$$x'(s) = y'(s) + \frac{h}{2}m(s, s_0)x(s_0) + h\sum_{j=1}^{n-1} m\big(s, s_j\big)x\big(s_j\big)$$
$$+\frac{h}{2}m(s, s_n)x(s_n) + \frac{h^2}{12}(l(s, s_0)x(s_0) + m(s, s_0)x(s_0)) \tag{5.42}$$
$$-\frac{h^2}{12}\big(l(s, s_n)x(s_n) + m(s, s_n)x'(s_n)\big)$$

By substituting $s = s_i$ in Eq. (5.42), we can write

$$
\begin{aligned}
x'(s_i) = y'(s_i) &+ \frac{h}{2}m(s_i, s_0)x(s_0) + h\sum_{j=1}^{n-1} m(s_i, s_j)x(s_j) \\
&+ \frac{h}{2}m(s_i, s_n)x(s_n) + \frac{h^2}{12}(l(s_i, s_0)x(s_0) + m(s_i, s_0)x(s_0)) \\
&- \frac{h^2}{12}\big(l(s_i, s_n)x(s_n) + m(s_i, s_n)x'(s_n)\big), i = 0, \ldots, n
\end{aligned}
\tag{5.43}
$$

Also, by substituting $i = 0$ and $i = n$ in Eq. (5.43), we obtain

$$
\begin{aligned}
x'(s_0) = y'(s_0) &+ \left(\frac{h}{2}m(s_0, s_0) + \frac{h^2}{12}l(s_0, s_0)\right)x(s_0) \\
&+ h\sum_{j=1}^{n-1} m(s_0, s_j)x(s_j) + \left(\frac{h}{2}m(s_0, s_n) - \frac{h^2}{12}l(s_0, s_n)\right)x(s_0) \\
&+ \frac{h^2}{12}\big(m(s_0, s_0)x'(s_0) - m(s_0, s_n)x'(s_n)\big)
\end{aligned}
\tag{5.44}
$$

And

$$
\begin{aligned}
x'(s_n) = y'(s_n) &+ \left(\frac{h}{2}m(s_n, s_0) + \frac{h^2}{12}l(s_n, s_0)\right)x(s_0) \\
&+ h\sum_{j=1}^{n-1} m(s_n, s_j)x(s_j) + \left(\frac{h}{2}m(s_n, s_n) - \frac{h^2}{12}l(s_n, s_n)\right)x(s_n) \\
&+ \frac{h^2}{12}\big(m(s_n, s_0)x'(s_0) - m(s_n, s_n)x'(x_n)\big)
\end{aligned}
\tag{5.45}
$$

From the Eqs. (5.41), (5.44) and (5.45), a system of $n + 3$ equations and $n + 3$ unknowns is obtained. Naturally, the system obtained from the case (2) is more accurate than the system obtained from the case (1), because the corrected trapezoidal method has been used instead of the composite trapezoidal method for obtaining it.

Chapter 6
Numerical Methods for Integral–Differential Equations

6.1 Introduction

Integro-differential equations are encountered in various fields of science. It plays an important role in many branches of linear and nonlinear functional analysis and their applications are in Theory of Science, Engineering and Social science.

In this chapter, we are going to discuss several methods to solve integro-differential equations numerically.

6.2 Integral–Differential Equations

In general, if the function x exists under the integral sign and its derivatives exist in the equation, we call it the integral differential equation. For example, the following equation

$$x'(s) = g(s, x(s)) + \lambda \int_a^s k(s, t, x(t))dt x(a) = \alpha \tag{6.1}$$

is the nonlinear first-order Volterra equation and

$$\begin{aligned} g(s) &= P(s)x''(s) + Q(s)x'(s) + R(s)x(s) + \lambda \int_a^b k(s, t)x(t)dt \\ e &= Cx(r) + Dx'(r) \end{aligned} \tag{6.2}$$

is the second-order Fredholm integral–differential equation where,

T. Allahviranloo and A. Esfandiari, *A Course on Integral Equations with Numerical Analysis*, Mathematical Engineering, https://doi.org/10.1007/978-3-030-85350-1_6

$$\begin{aligned} r &= (r_1, \ldots, r_m)^T, a \le r_i \le b \\ x(r) &= (x(r_1), \ldots, x(r_m))^T \\ x'(r) &= (x'(r_1), \ldots, x'(r_m))^T \end{aligned} \tag{6.3}$$

and for a p-order problem, the C and D matrices have $p \times m$ dimensions, and the matrix e has $p \times 1$ dimensions. In Eqs. (6.1) and (6.2), K, P, Q and R are known functions and x is an unknown function. If the kernel in the Eq. (6.2) is as $k(s, t, x(t))$, then the integral–differential equation will be nonlinear.

Equations (6.1) and (6.2) represent the boundary value equations. It is clear that there are no boundary conditions in integral equations, but these conditions appear in differential integral equations because they are necessary to prove the uniqueness of the answer, and this is the only difference between integral equations and integral–differential equations.

There are various solutions for solving integral–differential equations; however, we will only examine the expansion method in this book.

Consider the following integral–differential equation:

$$g(s) = x''(s) + \lambda \int_{a=0}^{b=1} k(s,t)x(t)dt \tag{6.4}$$

$$x(0) = x(1) = 0$$

Suppose that

$$x(s) = \sum\nolimits_{i=0}^{N} a_i h_i(s) \tag{6.5}$$

By substituting Eq. (6.4) in Eq. (6.5), we have

$$\begin{aligned} r_N(s) &= x_N''(s) + \lambda \int_0^1 k(s,t)x_N(t)dt - g(s) \\ &= \sum_{(i=0)}^{N} a_i \left(h_i''(s) + \lambda \int_0^1 k(s,t)h_i(t)dt \right) - g(s) \end{aligned} \tag{6.6}$$

The above equation can be solved by the Galerkin and collocation methods. As mentioned before, in the collocation method, we consider some points of the interval [0, 1] according to the number of unknowns in Eq. (6.6) such that

$$r_N(s_i) = 0, i = 0, 1, \ldots, N \tag{6.7}$$

In Eq. (6.6),

$$k_i(s) = \int_0^1 k(s,t)h_i(t)dt \tag{6.8}$$

is either calculated accurately or approximately.

6.2.1 Example

Consider the following equation

$$x''(s) - 60\int_0^1 (s-t)x(t)dt = s-2$$
$$x(0) = x(1) = 0$$

which has the real answer as $x(s) = s(s-1)^2$. The h_i's can be obtained such that they satisfy the boundary conditions. Suppose that

$$h_0(s) = s(s-1),\, h_1(s) = s^2(s-1)$$

where

$$\begin{cases} h_0(0) = h_0(1) = 0 \\ h_1(0) = h_1(1) = 0 \end{cases}$$

According to Eq. (6.5), we will have

$$x(s) = \sum_{i=0}^{1} a_i h_i(s) = a_0 h_0(s) + a_1 h_1(s)$$

It is clear that

$$h_0''(s) = 2,\, h_1''(s) = 6s-2$$

As a result

$$r_1(s) = \sum_{(i=0)}^{1} a_i(h_i''(s) - 60k_i(s)) - g(s)$$

$$= a_0(2 - 5(1 - 2s)) - (s - 2) + a_1(6s - 2 - (3 - 5s))$$

Therefore, according to Eq. (6.8), we have

$$k_0(s) = \int_0^1 (s - t)\, t\, (t - 1)\, dt = \frac{1}{12}(1 - 2s)$$

$$k_1(s) = \int_0^1 (s - t)\, t^2 (t - 1)\, dt = \frac{1}{60}(3 - 5s)$$

Now, according to Eq. (6.7), we assume that

$$s_0 = 0, s_1 = 1$$

So

$$r_1(0) = r_1(1) = 0$$

Therefore

$$\begin{cases} 3a_0 + 5a_1 = 2 \\ 7a_0 + 6a_1 = -1 \end{cases}$$

where by solving the above system, $a_0 = 1$ and $a_1 = 1$ are obtained. Then

$$x_1(s) = -s(s - 1) + s^2(s - 1) = s(s - 1)^2$$

is the real answer.

The number of h_i's can be considered to be large. For example, in the above example, we assume that:

$$h_0(s) = s(s - 1), h_1(s) = s^2(s - 1), h_2(s) = s^3(s - 1)$$

which are linearly independent and satisfy the boundary conditions. These functions are the same as s^2, s^3, s^4.

We now introduce the El-gendi method for solving integral–differential equations.

6.3 El-Gendi Method

We use this method for the integral–differential equations

$$g(s) = x'(s) + R(s)x(s) + \lambda \int_{-1}^{1} k(s,t)x(t)dt \tag{6.9}$$

By integrating both sides of Eq. (6.9), we have:

$$\int_{(-1)}^{s} g(t)dt = \int_{-1}^{s} x'(t)dt + \int_{-1}^{s} R(t)x(t)dt + \lambda \int_{-1}^{s} dt \int_{-1}^{1} k(t,u)x(u)du$$

Assuming $s = s_i$, we have

$$\int_{-1}^{s_i} g(t)dt = x(s_i) - x(-1) + \int_{-1}^{s_i} R(t)x(t)dt + \lambda \int_{-1}^{s_i} \int_{-1}^{1} k(t,u)x(u)dudt \tag{6.10}$$

where $x(-1) := e$ is defined. We consider

$$x(s) \simeq x_N(s) = \sum_{i=0}^{N}{}'' a_i T_i(s)$$

as a result

$$\int_{-1}^{1} \frac{x_N(s)T_j(s)}{\sqrt{1-s^2}} ds = \sum_{i=0}^{N}{}'' a_i \int_{-1}^{1} \frac{T_i(s)T_j(s)}{\sqrt{1-s^2}} ds = \frac{\pi}{2} a_j$$

where

$$a_i = \frac{2}{\pi} \int_{-1}^{1} \frac{x_N(s)T_i(s)}{\sqrt{1-s^2}} ds$$

If we use the closed Gauss–Chebyshev method to calculate the above integral, we will have

$$a_i \simeq \frac{2}{\pi} \cdot \frac{\pi}{N} \sum_{k=0}^{N}{}'' x_N\left(\cos\frac{k\pi}{N}\right)\cos\frac{ik\pi}{N}$$

where Chebyshev points of

$$s_k = \cos\frac{k\pi}{N}, k = 0, 1, \ldots, N$$

are nodal points.

We define

$$g_N(t) = \sum_{k=0}^{N}{}'' \alpha_k T_k(t) \tag{6.11}$$

By integrating both sides of Eq. (6.11), we have

$$\int_{-1}^{s} g_N(t)dt = \sum_{k=0}^{N}{}'' \alpha_k \int_{-1}^{s} T_k(t)dt$$

Given that $T_k(t) = \cos k\theta, \theta = \cos^{-1} t$ for $k > 1$, we will have

$$\begin{aligned}
\int_{(-1)}^{s} T_N(t)dt &= \int_{s'}^{\pi} \cos\, k\theta \cdot \sin\theta\, d\theta \\
&= \frac{1}{2}\int_{s'}^{\pi} (\sin(k+1)\theta - \sin(k-1)\theta)d\theta \\
&= \frac{1}{2}\left[\frac{\cos(k-1)\theta}{k-1} - \frac{\cos(k+1)\theta}{k+1}\right]_{s'}^{\pi} \\
&= \frac{1}{2}\left(\frac{(-1)^{k-1}}{k-1} - \frac{(-1)^{k+1}}{k+1} - \frac{T_{k-1}(s')}{k-1} + \frac{T_{k+1}(s')}{k+1}\right) \\
&= \frac{T_{k+1}(s')}{2(k+1)} - \frac{T_{k-1}(s')}{2(k-1)} + \frac{(-1)^{k+1}}{k^2-1}
\end{aligned} \tag{6.12}$$

where $s' = \cos^{-1} s$ is assumed and for $k = 0$ and $k > 1$, $\frac{(-1)^{k+1}}{k^2-1}$ is the integration constant value. For $k = 0$ and $k = 1$, we will have respectively

$$\int_{-1}^{s} T_0(t)dt = \int_{-1}^{s} dt = s + 1 = T_1(s) + 1$$

$$\int_{-1}^{s} T_1(t)dt = \int_{-1}^{s} dt = \frac{s^2}{2} - \frac{1}{2} = \frac{1}{4}(T_2(s) - 1) \tag{6.13}$$

In fact, if we integrate $g_N(t)$, the degree increases to $N + 1$. So

$$\int_{-1}^{s} g_N(t)dt = \sum_{r=0}^{N+1} \beta_r T_r(s) = \sum_{j=0}^{N} \alpha_j \int_{-1}^{s} T_j(t)dt \tag{6.14}$$

where

$$\beta_0 = \sum_{j=0, j\neq 1}^{N} {}'' \frac{(-1)^{j+1}\alpha_j}{j^2 - 1} - \frac{1}{4}\alpha_1$$

$$\beta_k = \frac{\alpha_{k-1} - \alpha_{k+1}}{2k}, \; k = 1, 2 \ldots, N - 2 \tag{6.15}$$

(Why?) and for $1 < k < N-1$, the coefficient of β_k is $T_k(s)$. For $k = N-1, N, N+1$, β_k's can be written as follows:

$$\beta_{N-1} = \frac{\alpha_{N-2} - \frac{1}{2}\alpha_N}{2(N-1)}$$

$$\beta_N = \frac{\alpha_{N-1}}{2N} \tag{6.16}$$

$$\beta_{N-1} = \frac{\frac{1}{2}\alpha_N}{2(N+1)}$$

According to Eq. (6.10), we have

$$\int_{-1}^{s_i} g(t)dt + e = x(s_i) + \int_{-1}^{s_i} R(t)x(t)dt$$

$$+ \lambda \int_{-1}^{s_i} \int_{-1}^{1} k(t, u)x(u)dudt \tag{6.17}$$

Using the expansion method, we define

$$g_N(t) = \sum_{i=0}^{N} {}'' \alpha_i T_i(t)$$

where

$$a_j = \frac{2}{N}\sum_{k=0}^{N}{}'' g_N(s_k)T_j(s_k),\ s_k = \cos\frac{k\pi}{N},\ j = 0, 1, \ldots, N \tag{6.18}$$

Using these α_j s, β_j can be calculated. So, for such a system, we will have

$$Gg_N = \bar{g}_N \tag{6.19}$$

The i-th equation of the system (6.19) is as follows

$$(\bar{g}_N)_i = \int_{-1}^{s_i} g_N(t)dt = \sum_{j=0}^{\infty} w_{ij} g_N(s_i),\ i = 0, \ldots, N$$

where

$$(G)_{ij} = w_{ij},\ g_N = \begin{bmatrix} g_N(s_0) \\ \vdots \\ g_N(s_N) \end{bmatrix}$$

Now, if $i = N$ is assumed, we will have

$$s_N = 1,\ (\bar{g}_N)_N = \int_{-1}^{1} g_N(t)dt$$

and finally

$$g_N(s_i) = g(s_i),\ i = 0, 1, \ldots, N \tag{6.20}$$

If $g_N(s_i)$ are given, we have α_i s, and therefore, we will have β_i s, so the system can be formed. Thus, it is enough to calculate $g_N(s_i)$.

6.3.1 Example

Consider Eq. (6.9). Suppose that $g(s)$ is approximated as follows.

$$g_2(s) = \sum\nolimits_{j=0}^{2}{}'' \alpha_j T_j(s) \tag{6.21}$$

So according to Eq. (6.14), we have

$$\int_{-1}^{s} g_2(t)dt = \sum_{r=0}^{3} \beta_r T_r(s)$$

where

$$\beta_0 = \frac{1}{2}\alpha_0 - \frac{1}{4}\alpha_1 - \frac{1}{6}\alpha_2$$
$$\beta_1 = \frac{1}{2}\alpha_0 - \frac{1}{4}\alpha_2$$
$$\beta_2 = \frac{1}{4}\alpha_1$$
$$\beta_3 = \frac{1}{12}\alpha_2$$

Also, according to Eqs. (6.18) and (6.20), we have

$$\alpha_j = {\sum_{k=0}^{2}}'' g(s_k)T_j(s_k),\ j = 0, 1, 2$$

So β_r s can be calculated. By substituting β_r in Eq. (6.18) and assuming $s_k = cos\frac{k\pi}{2}$, the following equation is obtained:

$$\int_{-1}^{s_i} g_2(t)dt = \sum_{j=0}^{2} w_{ij} g(s_j)$$

where

$$w_{00} = \frac{1}{3},\ w_{01} = \frac{4}{3},\ w_{02} = \frac{1}{3}$$
$$w_{10} = -\frac{1}{12},\ w_{11} = \frac{2}{3},\ w_{12} = \frac{5}{12}$$
$$w_{20} = w_{21} = w_{22} = 0$$

6.4 Fast Galerkin Method

Now, we use the fast Galerkin method to solve the first-order linear integral–differential equations:

$$\begin{aligned} g(s) &= Q(s)x'(s) + R(s)x(s) \\ &\quad + \lambda \int_{-1}^{1} k(s,t)x(t)dt, -1 \le s \le 1 \\ e &= c^T x(r) + d^T x'(r) \end{aligned} \tag{6.22}$$

where

$$\begin{aligned} r &= (r_1, \ldots, r_m)^T, -1 \le r_i \le 1 \\ c &= (c_1, \ldots, c_m)^T, d = (d_1, \ldots, d_m) \end{aligned} \tag{6.23}$$

and $x(r)$ and $x^{'}(r)$ are defined in Eq. (6.23).

Suppose that

$$\begin{aligned} x(s) &= \sum_{j=0}^{\infty}{}' a_j T_j(s) \\ x^{'}(s) &= \sum_{j=0}^{\infty}{}' a_j^{'} T_j(s) \\ Q(s) &= \sum_{j=0}^{\infty}{}' q_j T_j(s) \\ R(s) &= \sum_{j=0}^{\infty}{}' r_j T_j(s) \\ g(s) &= \sum_{j=0}^{\infty}{}' g_j T_j(s) \end{aligned} \tag{6.24}$$

If we substitute Eq. (6.24) in Eq. (6.22) and integrate them, we will have

$$\sum_{k=0}^{\infty} g_k \int_{-1}^{1} \frac{T_k(s)T_i(s)}{\sqrt{1-s^2}} ds \sum_{j=0}^{\infty}{}' a_j^{'} \sum_{k=0}^{\infty}{}' q_k \int_{-1}^{1} \frac{T_k(s)T_j(s)T_i(s)}{\sqrt{1-s^2}} ds$$

$$
+\sum_{j=0}^{\infty}{}' a_j \sum_{k=0}^{\infty}{}' r_k \int_{-1}^{1} \frac{T_k(s)T_j(s)T_i(s)}{\sqrt{1-s^2}} ds
$$

$$
+\lambda \int_{-1}^{1} \frac{k(s,t)T_j(s)T_i(s)}{\sqrt{1-s^2}} ds \tag{6.25}
$$

According to $T_i(s) = \cos i\theta$ and $s = cos\theta$, we have

$$
\begin{aligned}
T_i(s)T_j(s) &= \cos\ i\ \theta \cdot \cos\ j\ \theta \\
&= \frac{1}{2}(\cos(i+j)\theta - \cos(i-j)\theta) \\
&= \frac{1}{2}\left(T_{i+j}(s) + T_{|i-j|}(s)\right)
\end{aligned} \tag{6.26}
$$

And thus, according to Eq. (6.26),
we obtain

$$
\int_{-1}^{1} \frac{T_k(s)T_j(s)T_i(s)}{\sqrt{1-s^2}} ds = \frac{1}{2}\int_{-1}^{1} \frac{T_k(s)T_{i+j}(s)}{\sqrt{1-s^2}} ds
$$

$$
+\frac{1}{2}\int_{-1}^{1} \frac{T_k(s)T_{i-j}(s)}{\sqrt{1-s^2}} ds
$$

$$
= \begin{cases} \pi, i = j = k = 0 \\ \frac{\pi}{2}\delta_{ij}, k = 0, i + j > 0 \\ \frac{\pi}{4}\left(\delta_{k,i+j} + \delta_{k,|i-j|}\right), k > 0 \end{cases}
$$

According to Eq. (6.25), the following system can be formed

$$
Q\underline{a}' + (R + \lambda B)\underline{a} = \underline{g} \tag{6.27}
$$

where

$$
Q_{ij} = \frac{q_{i+j} + q_{|i-j|}}{2}
$$

$$
R_{ij} = \frac{r_{i+j} + r_{|i-j|}}{2}
$$

$$
B_{ij} = \frac{2}{\pi}\int_{-1}^{1}\int_{-1}^{1} \frac{k(s,t)T_j(s)T_i(s)}{\sqrt{1-s^2}} dsdt, i = 0, \ldots, j = 1, \ldots
$$

and for $i \geq 0$ and $j = 0$, we define

$$Q_{i0} = \frac{q_i}{2},\ R_{i0} = \frac{r_i}{2}$$

$$B_{i0} = \frac{1}{\pi}\int_{-1}^{1}\int_{-1}^{1}\frac{k(s,t)T_i(s)}{\sqrt{1-s^2}}dsdt$$

Since $Q(s)$ are given, we can obtain q_j s by the closed Gauss–Chebyshev method. The coefficients a'_j are related to a_j with the following equation:

$$a_j = \frac{a'_{j-1} - a'_{j+1}}{2j},\ j = 1, 2, \ldots$$

Or it can be written as:

$$a^{(1)} = Aa' \tag{6.28}$$

where

$$A_{ij} = \begin{cases} \frac{1}{2(j+1)}, & j = i \geq 0 \\ \frac{-1}{2(j-1)}, & j = i+2 \\ 0 o.w \end{cases}$$

$$a^{(1)} = (a_1, a_2, \ldots)^T$$
$$a' = \left(a'_0, a'_1, \ldots\right)^T$$

Therefore, the system (6.28) can be written as follows

$$\begin{bmatrix} \frac{1}{2} & 0 & -\frac{1}{2} & & & & \\ & \frac{1}{4} & 0 & -\frac{1}{4} & & & \\ & & \frac{1}{6} & 0 & -\frac{1}{6} & & \\ & & \ddots & \ddots & \ddots & & \\ & & & \frac{1}{2(j+1)} & 0 & -\frac{1}{2(j+1)} & \\ & & & & \ddots & \ddots & \ddots \end{bmatrix} \begin{bmatrix} a'_0 \\ a'_1 \\ \vdots \\ a'_N \\ \vdots \end{bmatrix} = \begin{bmatrix} a_1 \\ a_2 \\ \vdots \\ a_N \\ \vdots \end{bmatrix}$$

where

$$\frac{a'_{j-1}}{2j} - \frac{a'_{j+1}}{2j} + a'_j \times 0 = a_j$$

a_0 is also obtained from the boundary conditions of (6.25), so that if we write the boundary conditions as follows

$$\sum_{i=1}^{m}\big(c_i x(r_i) + d_i x'(r)\big) = e$$

by substituting the Chebyshev expansion of $x(r_i)$ and $x'(r_i)$ in the above equation, we will have:

$$\sum_{i=1}^{m}\left(c_i \sum_{j=0}^{\infty}{}' a_j T_j(r_i) + d_i \sum_{j=0}^{\infty}{}' a'_j T_j(r_i)\right) = e$$

Or

$$c^T T a + d^T T a' = e \tag{6.29}$$

where

$$T_{ij} = T_j(r_i),\ i = 1, \ldots, m,\ j = 1, 2, \ldots$$
$$T_{i0} = \frac{1}{2},\ i = 1, \ldots, m$$

We can also express $c^T T a$ as follows

$$c^T T a = (cI)a_0 + c^T T^{(1)} a^{(1)} \tag{6.30}$$

where

$$I = \left(\frac{1}{2}, \ldots, \frac{1}{2}\right)^T,\ T^{(1)} = T_{j,j+1},\ i = 1, \ldots, m,\ j = 1, 2, \ldots$$

By rearranging Eq. (6.30) and according to Eq. (6.28), we obtain

$$a_0 = \frac{e - \big(d^T + c^T T^{(1)} A\big)a'}{\Delta_1} \tag{6.31}$$

where $\Delta_1 = c^T I \neq 0$. Now if we put

$$\mu = \frac{e}{\Delta_1},\ k^T = -\frac{d^T + c^T T^{(1)} A}{\Delta_1}$$

By combining Eqs. (6.28) and (6.31), we will have

$$a = A'a' + \mu \tag{6.32}$$

where

$$A^{'} = \begin{bmatrix} k' \\ A \end{bmatrix}, \mu = \begin{bmatrix} \mu \\ 0 \end{bmatrix}$$

We now substitute Eq. (6.32) in Eq. (6.27), and finally, the following system is obtained

$$\left(Q + (R + \lambda B)A^{'}\right)a^{'} = g_1$$

where

$$g_1 = g - (R - \lambda B)\mu$$

And by solving this system, we obtain a'.

Chapter 7
Introduction to Interval Integral Equations

Check for updates

7.1 Introduction

In this chapter, we are going to consider the interval Fredholm integral equations in which the deriving terms are an interval then the solution must be a function. In these equations, it is supposed that the kernel is a real-valued function and only the deriving terms are an interval function.

7.2 Interval Fredholm Integral Equations

Since solving the interval integral equations by numerical methods leads to solving a linear system of equations, then the short description about these systems is brought here.

7.2.1 *Definition-Dual Interval System*

The $n \times n$ dual linear system

$$\begin{cases} a_{11}\tilde{x}_1 + \cdots + a_{1n}\hat{x}_n = \tilde{y}_1 + b_{11}\tilde{x}_1 + \cdots + b_{1n}\tilde{x}_n, \\ a_{21}\tilde{x}_1 + \cdots + a_{2n}\hat{x}_n = \tilde{y}_2 + b_{21}\tilde{x}_1 + \cdots + b_{2n}\tilde{x}_n \\ \\ a_{n1}\tilde{x}_1 + \cdots + a_{nn}\hat{x}_n = \tilde{y}_1 + b_{n1}\tilde{x}_1 + \cdots + b_{nn}\tilde{x}_n. \end{cases} \tag{7.1}$$

where the coefficient matrix $A = (a_{ij})$ and $B = \left(b_{ij}\right), 1 \leq i, j \leq n$ are real $n \times n$ matrices, $\tilde{x}^t = (\tilde{x}_1, \ldots, \tilde{x}_n)$ be a $n \times 1$ vector of interval numbers $\tilde{x}_j$ and $\tilde{y}^t = (\tilde{y}_1, \ldots, \tilde{y}_n)$ be a $n \times 1$ vector of interval numbers in which $\tilde{y}_i$ is called a dual interval linear system.

T. Allahviranloo and A. Esfandiari, *A Course on Integral Equations with Numerical Analysis*, Mathematical Engineering, https://doi.org/10.1007/978-3-030-85350-1_7

7.2.2 Definition

An interval number vector $(\tilde{x}_1, \tilde{x}_2, \ldots, \tilde{x}_n)^t$ is given by components as

$$\tilde{x}_i = \left[\underline{x}_i, \overline{x}_i\right], 1 \leq i \leq n$$

is called a solution of the interval linear system (7.1) if

$$\begin{cases} \underline{\sum_{j=1}^{n} a_{ij}x_j} = \sum_{j=1}^{n} \underline{a_{ij}x_j} = \underline{y}_i = \underline{\sum_{j=1}^{n} b_{ij}x_j} = \sum_{j=1}^{n} \underline{b_{ij}x_j}, \\ \overline{\sum_{j=1}^{n} a_{ij}x_j} = \sum_{j=1}^{n} \overline{a_{ij}x_j} = \overline{y}_i = \overline{\sum_{j=1}^{n} b_{ij}x_j} = \sum_{j=1}^{n} \overline{b_{ij}x_j} \end{cases}$$

If, for a particular i, $a_{ij} \geq 0$ and $b_{ij} \geq 0$, $1 \leq j \leq n$, we simply get

$$\sum_{j=1}^{n} a_{ij}\underline{x}_j = \underline{y}_i + \sum_{j=1}^{n} b_{ij}\underline{x}_j, \quad \sum_{j=1}^{n} a_{ij}\overline{x}_j = \overline{y}_i + \sum_{j=1}^{n} b_{ij}\overline{x}_j$$

The following theorem guarantees the existence of an interval solution for general case. Consider the dual interval linear system (7.1), and transform its $n \times n$ coefficient matrix A and B into $(2n) \times (2n)$ matrices as:

$$\begin{cases} s_{11}\underline{x}_1 + \cdots + s_{1n}\underline{x}_n + s_{1,n+1}(-\bar{x}_1) + \cdots + s_{1,2n}(-\bar{x}_n) = y_1 + t_{11}\underline{x}_1 + \cdots + t_{1n}\underline{x}_n + t_{1,n+1}(-\bar{x}_1) + \cdots + t_{1,2n}(-\bar{x}_1) \\ \vdots \\ s_{n1}\underline{x}_1 + \cdots + s_{nn}\underline{x}_n + s_{n,n+1}(-\bar{x}_1) + \cdots + s_{n,2n}(-\bar{x}_n) = y_n + t_{n1}\underline{x}_1 + \cdots + t_{nn}\underline{x}_n + t_{n,n+1}(-\bar{x}_1) + \cdots + t_{n,2n}(-\bar{x}_1) \\ s_{n+1,1}\underline{x}_1 + \cdots + s_{n+1,n}\underline{x}_n + s_{n+1,n+1}(-\bar{x}_1) + \cdots + s_{n+1,2n}(-\bar{x}_n) = -\bar{y}_1 + t_{n+1,1}\underline{x}_1 + \cdots + t_{n+1,n}\underline{x}_n + t_{n+1,n+1}(-\bar{x}_1) + \cdots + t_{n+1,2n}(-\bar{x}_n) \\ \vdots \\ s_{2n,1}\underline{x}_1 + \cdots + s_{2n,n}\underline{x}_n + s_{2n,n+1}(-\bar{x}_1) + \cdots + s_{2n,2n}(-\bar{x}_n) = -\bar{y}_1 + t_{2n,1}\underline{x}_1 + \cdots + t_{2n,n}\underline{x}_n + t_{2n,n+1}(-\bar{x}_1) + \cdots + t_{2n,2n}(-\bar{x}_n) \end{cases}$$

where s_{ij} and t_{ij} are determined as follows:

$$a_{ij} \geq 0 \rightarrow s_{ij} = a_{ij}, s_{i+n,j+n} = a_{ij},$$

$$a_{ij} < 0 \rightarrow s_{i,j+n} = -a_{ij}, s_{i+n,j} = -a_{ij},$$

$$b_{ij} \geq 0 \rightarrow t_{ij} = b_{ij}, t_{i+n,j+n} = -b_{ij},$$

$$b_{ij} < 0 \rightarrow t_{i,j+n} = -b_{ij}, t_{i+n,j} = -b_{ij}$$

and s_{ij} any t_{ij} those which are not determined by (7.2) are zero. Using the matrix notation, we get

$$SX = Y + TX \tag{7.3}$$

therefore, we have:

$$(S - T)X = Y \tag{7.4}$$

where $S = (s_{ij}) \geq 0$ and $T = \left(t_{ij}\right) \geq 0$, $1 \leq i, j \leq 2n$, and

$$X = \begin{bmatrix} \underline{x}_1 \\ \vdots \\ \underline{x}_n \\ -\overline{x}_1 \\ \vdots \\ -\overline{x}_n \end{bmatrix}, Y = \begin{bmatrix} \underline{y}_1 \\ \vdots \\ \underline{y}_n \\ -\overline{y}_1 \\ \vdots \\ -\overline{y}_n \end{bmatrix} \tag{7.5}$$

For example:
Consider the dual interval linear system

$$\begin{cases} \tilde{x}_1 + \tilde{x}_2 = \tilde{y}_1 + 2\tilde{x}_1 + \tilde{x}_2 \\ \tilde{x}_1 + 2\tilde{x}_2 = \tilde{y}_2 + \tilde{x}_1 - 2\tilde{x}_2 \end{cases} \tag{7.6}$$

Let $\underline{y}_1 = 0, \overline{y}_1 = 2$ and $\underline{y}_2 = 4, \overline{y}_2 = 7$ the extended 4×4 matrices are

$$S = \begin{bmatrix} 1 & 0 & 0 & 1 \\ 1 & 2 & 0 & 0 \\ 0 & 1 & 1 & 0 \\ 0 & 0 & 1 & 2 \end{bmatrix}, T = \begin{bmatrix} 2 & 1 & 0 & 0 \\ 1 & 0 & 0 & 2 \\ 0 & 0 & 2 & 1 \\ 0 & 2 & 1 & 0 \end{bmatrix}$$

and

$$Y = \begin{bmatrix} 0 \\ 4 \\ -2 \\ -7 \end{bmatrix}, X = \begin{bmatrix} \underline{x}_1 \\ \underline{x}_2 \\ -\overline{x}_1 \\ -\overline{x}_2 \end{bmatrix}$$

We obtain that the system (7.6) is equivalent to the dual system of equations

$$SX = Y + TY$$

Consequently,

$$\begin{bmatrix} 1 & 0 & 0 & 1 \\ 1 & 2 & 0 & 0 \\ 0 & 1 & 1 & 0 \\ 0 & 0 & 1 & 2 \end{bmatrix} \begin{bmatrix} \underline{x}_1 \\ \underline{x}_2 \\ -\bar{x}_1 \\ -\bar{x}_2 \end{bmatrix} = \begin{bmatrix} 0 \\ 4 \\ -2 \\ -7 \end{bmatrix} + \begin{bmatrix} 2 & 1 & 0 & 1 \\ 1 & 0 & 0 & 2 \\ 0 & 0 & 2 & 1 \\ 0 & 2 & 1 & 0 \end{bmatrix} \begin{bmatrix} \underline{x}_1 \\ \underline{x}_2 \\ -\bar{x}_1 \\ -\bar{x}_2 \end{bmatrix}$$

Also,

$$(S - T)X = Y$$

The structure S of and T implies that:

$$S = \begin{pmatrix} C & D \\ D & C \end{pmatrix}, T = \begin{pmatrix} E & F \\ F & E \end{pmatrix}$$

where C and E contain the positive entries of A and B, respectively, and D and F the absolute values of the negative entries of A and B, i.e., $A = C - D$ and $B = E = F$. Therefore,

$$S - T = \begin{pmatrix} C - E & D - F \\ D - F & C - E \end{pmatrix}$$

7.2.3 Theorem

The matrix $S - T$ is nonsingular if and only if the matrix $(C + D) - (E + F)$ and $(C + F) - (E + D)$ are both nonsingular.

Proof. Assuming that is nonsingular, then the solution vector is found as (7.4).

$$X = (S - T)^{-1}Y \tag{7.7}$$

If $(S - T)^{-1}$ exists, it must have the same structure as S, i.e.,

$$(S - T)^{-1} = \begin{pmatrix} G & H \\ H & G \end{pmatrix}$$

and

$$G = \frac{1}{2}\left[((C + D) - (E - F))^{-1} + ((C + F) - (E + D))^{-1}\right]$$

$$H = \frac{1}{2}\left[((C + D) - (E - F))^{-1} - ((C + F) - (E + D))^{-1}\right]$$

7.2.4 Remark

The unique solution X of Eq. (7.7) is an interval vector for arbitrary Y if and only if $(S - T)^{-1}$ is nonnegative, i.e.,

$$\left((S - T)^{-1}\right)_{ij} \geq 0, 1 \leq i \leq 2n, 1 \leq j \leq 2n.$$

7.2.5 Definition: The Interval Number Vector

Let $X = \left[\underline{x}_i, \overline{x}_i\right]$, $1 \leq i \leq n$ denotes the unique solution of $SX = Y + TX$, if $\underline{y}, \overline{y}$ are linear functions, then the interval number vector $U = \left[\underline{u}_i, \overline{u}_i\right]$, $1 \leq i \leq n$ defined by

$$\underline{u}_i = min\left[\underline{x}_i, \overline{x}_i\right]$$

$$\overline{u}_i = max\left[\underline{x}_i, \overline{x}_i\right]$$

is called the interval solution of $SX = Y + TX$. If $\left[\underline{x}_i, \overline{x}_i\right]$, $1 \leq i \leq n$, are all interval numbers, then $\underline{u}_i = x_i, \overline{u}_i = \overline{x}_i$, and then U is called a **strong interval solution**. Otherwise, U is a **weak interval solution**.

Next, we will define a metric D on the set of all interval numbers.

7.2.6 Definition

For arbitrary interval numbers $\tilde{u} = \left[\underline{u}, \overline{u}\right]$, $\tilde{v} = \left[\underline{v}, \overline{v}\right]$ the quantity

$$D\left[\tilde{u}, \tilde{v}\right] = max\left[\left|\underline{u} - \underline{v}\right|, \left|\overline{u} - \overline{v}\right|\right]$$

is the distance between $\tilde{u}$, $\tilde{v}$.

7.2.7 *Definition*

Let $\tilde{U} = \left[\tilde{u}_1, \ldots, \tilde{u}_n\right] and\ \tilde{V} = \left[\tilde{v}_1, \ldots, \tilde{v}_n\right]$, D is the distance between these two interval number vectors is denoted by $D[\tilde{U}, \tilde{V}]$ and defined as

$$D\left[\tilde{U}, \tilde{V}\right] = \begin{bmatrix} D[\tilde{u}_1, \tilde{v}_1] \\ \vdots \\ D[\tilde{u}_m, \tilde{v}_n] \end{bmatrix}$$

7.3 Interval Fredholm Integral Equation

The Fredholm integral equation of the second kind is

$$x(s) = f(s) + \lambda \int_a^b k(s,t)x(t)dt \tag{7.11}$$

where $\lambda > 0$, $k(s, t)$ is an arbitrary kernel function over the square $a \leq s, t \leq b$ and $f(t)$ is a function of$t : a \leq t \leq b$. If $f(t)$ is a real function, then the solution of Eq. (7.11) is real as well. However, if $f(t)$ is an interval function, this equation my only possesses interval solution. Therefore, we have

$$\tilde{x}(s) = \tilde{f}(s) + \lambda \int_a^b k(s,t)\tilde{x}(t)dt \tag{7.12}$$

We consider now the numerical solution of interval Fredholm integral equations of the second kind Eq. (7.12), which we write in the form of system:

$$\tilde{x} = \tilde{f} + \lambda K \tilde{x} \tag{7.13}$$

Considering the expansion method, the exact solution of integral equation Eq. (7.12) is:

$$\tilde{x}(s) = \sum_{i=1}^{\infty} \tilde{a}_i h_i(s) \tag{7.14}$$

in truncated form

$$\tilde{x}(s) \approx \tilde{x}_n(s) = \sum_{i=1}^{n} \tilde{a}_i h_i(s) \tag{7.15}$$

where the set $\{h_i\}$ is complete and orthogonal in $\ell^2(a, b)$. For finding approximation solution, we must indicate coefficient $\tilde{a}_i$.

From Eq. (7.15), we obtain

$$\sum_{j=1}^{n} \tilde{a}_j h_j(s) = \tilde{f}(s) + \lambda \sum_{j=1}^{n} \tilde{a}_j \int_a^b k(s,t) h_j(t) dt \tag{7.16}$$

We have n unknown parameters in the form $\tilde{a}_1, \tilde{a}_2, \ldots, \tilde{a}_n$ which for finding them, we need to n equation, so by using n points $s_1, s_2, \ldots, s_n$ in interval $[a, b]$:

$$\sum_{j=1}^{n} h_j(s_j) \tilde{a}_j = \tilde{f}(s_j) + \lambda \sum_{j=1}^{n} \tilde{a}_j \int_a^b k(s_j, t) h_j(t) dt \tag{7.17}$$

therefore, we have:

$$A\tilde{a} = \tilde{f} + B\tilde{a} \tag{7.18}$$

where the coefficients matrix $A = (a_{ij})$, $1 \leq i, j \leq n$, and $B = (b_{ij})$, $1 \leq i, j \leq n$, are real and $\tilde{f} = (\tilde{f}_i)$, $1 \leq i \leq n$, is an arbitrary interval number vector, where

$$a_{ij} = h_j(s_i), b_{ij} = \lambda \int_a^b k(s_i, t) h_j(t) dt, i, j = 1, \ldots, n$$

7.3.1 Residual Minimization Method

The simplest method conceptually again appeals to approximation theory. We write the integral equation in the form (again we set $\lambda = 1$)

$$L\tilde{x} = \tilde{f}, L = I - K \tag{7.19}$$

and introduce the residual function r_n and error function ε_n

$$r_n = D(\tilde{f}, L\tilde{x}_n) \tag{7.20}$$

$$\varepsilon_n = D(\tilde{x}, \tilde{x}_n) \tag{7.21}$$

where D is any distance function and to compute r_n requires no knowledge of $\tilde{x}$ but, since $\dot{D}\left(\tilde{f}, L\tilde{x}\right) = 0$, we have the identify:

$$r_n = D\left(\tilde{f}, L\tilde{x}_n\right) - D\left(\tilde{f}, L\tilde{x}\right) = LD(\tilde{x}_n, \tilde{x}) = L\varepsilon_n. \tag{7.22}$$

From (7.19) and (7.22), we have at once

$$\|r_n\| \leq (1 + \|K\|)\|\varepsilon_n\| \tag{7.23}$$

That is,

$$\|\varepsilon_n\| \geq \frac{\|r_n\|}{1+K} \tag{7.24}$$

Thus, a small residual is necessary condition for a small error. We would rather have an upper bound on ε_n of course; this is harder to provide in general, and we content ourselves for now with the following.

We rewrite (7.22) as

$$\varepsilon_n = r_n + K\varepsilon_n$$

whence

$$\|\varepsilon_n\| \leq \|r_n\| + \|K\|.\|\varepsilon_n\| \tag{7.25}$$

and hence if

$$\|\varepsilon_n\| < 1$$

$$\|\varepsilon_n\| \leq \frac{\|r_n\|}{1-K} \tag{7.26}$$

Consider the interval integral Eq. (7.12) where

$$\underline{f}(s) = 0,$$

$$\overline{f}(s) = 4s^3$$

and kernel

$$k(s,t) = s + 1, -1 \leq s, t \leq 1$$

and $a = 1, b = -1$. The exact solution in this case is given by

$$\underline{x}(s) = 0,$$

$$\overline{x}(s) = 4s^3.$$

and

$$s_1 = -1, s_2 = 1$$

From Eqs. (7.17) and (7.18):

$$\begin{bmatrix} 1 & -1 \\ 1 & 1 \end{bmatrix} \begin{bmatrix} \tilde{a}_1 \\ \tilde{a}_2 \end{bmatrix} = \begin{bmatrix} \tilde{f}(s_1) \\ \tilde{f}(s_2) \end{bmatrix} + \begin{bmatrix} 0 & 0 \\ 4 & 0 \end{bmatrix} \begin{bmatrix} \tilde{a}_1 \\ \tilde{a}_2 \end{bmatrix}$$

the extended 4×4 matrices are

$$S = \begin{bmatrix} 1 & 0 & 0 & 1 \\ 1 & 1 & 0 & 0 \\ 0 & 1 & 1 & 0 \\ 0 & 0 & 1 & 1 \end{bmatrix}, T = \begin{bmatrix} 0 & 0 & 0 & 0 \\ 4 & 0 & 0 & 0 \\ 0 & 0 & 0 & 0 \\ 0 & 0 & 4 & 0 \end{bmatrix}$$

and the solution of Eq. (7.7) is:

$$\begin{bmatrix} \underline{a}_1 \\ \underline{a}_2 \\ -\overline{a}_1 \\ -\overline{a}_2 \end{bmatrix} = (S - T)^{-1} F = \begin{bmatrix} 1 \\ 3 \\ 1 \\ 1 \end{bmatrix} \tag{7.27}$$

The fact that after solving this system it is found that $\tilde{a}_1 = [\underline{a}_1, \overline{a}_1]$ and $\tilde{a}_2 = [\underline{a}_2, \overline{a}_2]$ are not interval numbers. Therefore, the interval solution of Eq. (7.27) is a weak interval solution.

References

1. S. Abbasbandy, T. Allahviranloo, Numerical solution of fuzzy differential equations. Math. Comput. Appl. **7**(1), 41–52 (2002)
2. S. Abbasbandy, T. Allahviranloo, Numerical solution of fuzzy differential equations, numerical solutions of fuzzy differential equations by Taylor method. Comput. Methods Appl. Math. **2**(2), 113–124 (2002)
3. S. Abbasbandy, T. Allahviranloo, O. Lopez-Pouso, J.J. Nieto, Numerical methods for fuzzy differential inclusions. J. Comput. Math. Appl. **48**, 1633–1641 (2004)
4. S. Abbasbandy, T. Allahviranloo, The Adomian decomposition method applied to the fuzzy system of Fredholm integral equations of the second kind. Int. J. Uncertainty, Fuzziness Knowl Based Syst. **14**(01), 101–110 (2006)
5. T. Allahviranloo, E. Ahmady, N. Ahmady, K. Shams Alketaby, Block Jacobi two-stage method with Gauss–Sidel inner iterations for fuzzy system of linear equations. Appl. Math. Comput. Appl. Math. Comput. **175**(2), 1217–1228 (2006)
6. T. Allahviranloo, M. Adabitabar Firozja, Note on "Trapezoidal approximation of fuzzy numbers". Fuzzy Sets Syst. 158(7), 755–756 (2007)
7. T. Allahviranloo, Numerical methods for fuzzy system of linear equations. Appl. Math. Comput. **155**(2), 493–502 (2004)
8. T. Allahviranloo, S. Abbasbandy, H. Rouhparvar, The exact solutions of fuzzy wave-like equations with variable coefficients by a variational iteration method. Appl. Soft Comput. **11**(2), 2186–2192 (2011)
9. T. Allahviranloo, Successive over relaxation iterative method for fuzzy system of linear equations. Appl. Math. Comput. Appl. Math. Comput. **162**(1), 189–196 (2005)
10. T. Allahviranloo, S. Salahshour, Euler method for solving hybrid fuzzy differential equation. Soft. Comput. **15**, 1247–1253 (2011)
11. S. D. Conte, C. de Boor, *Elementary Numerical Analysis an Algorithmic Approach* (McGraw-Hill, New York, 1980)
12. D. Dubois, H. Prade, Towards fuzzy differential calculus. Fuzzy Sets Syst. **8**, 1–7, 105–116, 225–233 (1982)
13. D. Kincaid, W. Cheney, *Numerical Analysis: Mathematics of Scientific Computing* (Brooks/Cole, 1996)
14. P. J. Davis, *Interpolation and Approximation* (Dover Publications, New York, 1975)
15. L. M. Delves, J. L. Mohamed, *Computational Methods for Integral Equations* (Cambridge University Press, 1985)
16. W. Gautschi, Gauss Kronrod Quadrature—A Survey. Numer. Methods Approx. **III**, 39–66 (1988)
17. W. Gautschi, *Adaptive Quadrature—Revisited* (1998)

T. Allahviranloo and A. Esfandiari, *A Course on Integral Equations with Numerical Analysis*, Mathematical Engineering, https://doi.org/10.1007/978-3-030-85350-1

18. G. Hammerlin, K.H. Hoffmann, *Numerical Mathematics* (Springer, New York, 1991)
19. G. Hammerlin, K.H. Hoffmann, *Numerical Mathematics* (Springer, New York, 1991)
20. F.B. Hildebrand, *Introduction to Numerical Analysis* (Dover Publications, New York, 1974)
21. E. Isaacson, H. BishopKeller, *Analysis of Numerical methods* (Wiley, New York, 1966)
22. E. Isaacson, H. Bishop Keller, *Analysis of Numerical Methods* (Wiley, New York, 1966)
23. J.J. Buckley, Fuzzy eigenvalues and input–output analysis. FSS **34**, 187–195 (1990)
24. J. M. Ortega, *Numerical Analysis a Second Course* (Siam, 1990)
25. K. Atkinson, *An Introduction to Numerical Analysis*, 2nd edn (Wiley, 1989)
26. D. Kincaid, W. Cheney, *Numerical Analysis: Mathematics of Scientific Computing* (Brooks-Cole, 2002)
27. D. Kincaid, W. Cheney, *Numerical Analysis: Mathematics of Scientific Computing* (Brooks-Cole, 2002)
28. A. S. Kronrod, *Nodes and Weights of Quadrature Formula* (New York, 1965)
29. L.A. Zadeh, The concept of a linguistic variable and its application to approximate reasoning. Inform. Sci. **8**, 199–249 (1975)
30. J. N. Lyness, Notes on the adaptive Simpson quadrature routine. J. Assoc. Comput. Mach. **16**, 483–492 (1969)
31. J. N. Lyness, Notes on the adaptive Simpson quadrature routine. J. Assoc. Comput. Mach. **16**, 483–492 (1969)
32. M. Friedman, M. Ming, A. Kandel, Fuzzy linear systems. FSS **96**, 201–209 (1998)
33. M. Friedman, M. Ma, A. Kandel, Numerical solutions of fuzzy differential equations and integral equations. Fuzzy Sets Syst. **106**, 35–48 (1999)
34. M. Friedman, M. Ming, A. Kandel, Duality in fuzzy linear systems. Fuzzy Sets Syst. **109**, 55–58 (2000)
35. C. D. Meyer, *Matrix Analysis and Applied Linear Algebra* (Siam, 2000)
36. J. M. Ortega, *Numerical Analysis: A Second Course* (Siam, 1990)
37. P. Diamond, Fuzzy least squares. Inform. Sci. **46**, 144–157 (1988)
38. G.M.M. Phillips, P.J. Taylor, *Theory and Applications of Numerical Analysis* (Elsevier Science and Technology Books, 1996)
39. R. Goetschel, W. Voxman, Elementary calculus. Fuzzy Sets Syst. **18**, 31–43 (1986)
40. A. Ralston, P. Rrabinowitz, *A First Course in Numerical Analysis* (McGraw-Hill, New York, 1982)
41. A. Ralston, P. Rrabinowitz, *A First Course in Numerical Analysis* (McGraw-Hill, New York, 1982)
42. S.L. Chang, L.A. Zadeh, On fuzzy mapping and control, IEEE Trans. Syst. Man Cyb. **2**, 30–34 (1972)
43. D. Sheen, *Introduction to Numerical Analysis* (Seoul National University, 2007)
44. J. Stoer, R. Bulirsch, *Introduction to Numerical Analysis* (Springer, New York, 2002)
45. J. Stoer, R. Bulirsch, *Introduction to Numerical Analysis* (Springer, New York, 2002)
46. G. Szego, Orthogonal polynomials. Am. Math. (1975)
47. G. Szego, Uber gewisse orthogonal polynomials. Math. Ann. (1935)
48. W. Cong-Xin, M. Ming, Embedding problem of fuzzy number space: Part I. Fuzzy Sets Syst. **44**, 33–38 (1991)
49. C.J. Zarowski, *An Introduction to Numerical Analysis for Electrical and Computer Engineers* (Wiley , New Jersey, 2004)
50. Z. Gouyandeh, T. Allahviranloo, S. Abbasbandy, A. Armand, A fuzzy solution of heat equation under generalized Hukuhara differentiability by fuzzy Fourier transform. Fuzzy Sets Syst. **309**(15), 81–97 (February 2017)
51. A. Ralston, P. Rabinowitz, *A First Course in Numerical Analysis* (McGraw-Hill, 1978)
52. J. Stoer, R. Bulirsch, *Introduction to Numerical Analysis* (Springer, 1993)